Green Planet

All about green technologies and science

By Prasun Barua

About

Welcome to Green Planet! First of all, I would like to thank you for visiting and reading Green Planet. This is Prasun Barua, the Author of Green Planet. I am an Engineer (Electrical & Electronic) and Member of the European Energy Centre (EEC). Green Planet is all about green technologies and science. The motivation of Green Planet is to create a green and healthy planet utilizing green technologies and science as well as our awareness. Green Planet contains various types of topics categorized to Green Energy, Green Technology, Green Products, Green Finance, Green IT, Green Jobs, Green Nature and Health. It will be a great pleasure if Green Planet keeps contribution in your daily life.

Our planet has many significant contributions in our daily life. We are benefiting from our planet in various ways. Green nature is one of the great creations of our planet. For example, trees provide us oxygen, so that we can live. They also provide us fruit which contains various types of vitamins, minerals and nutrition. Vegetables are also a good source of vitamins, minerals and nutrition. These fruits and vegetables are essential for our health. They protect us from various types of diseases. Green forests protect us from many natural disasters. Green technologies and science are also very significant for us. We can generate power with the help of renewable energy resources like the ray of sun, force of air, water, biomass, internal heat of the earth and the tidal wave of the ocean. These technologies are called green technologies. They do not pollute our environment during generating power. They contribute us to reduce the emission of carbon. They are very environmentally friendly technologies indeed.

Every day, we are damaging the nature of our planet consciously or unconsciously. For example, we are cutting trees randomly for our daily livelihood. Cutting the huge amount of trees can create imbalance of our ecosystem which impacts on human being and animals. But, we are not creating a huge tree plantation as per our requirement. Moreover, we are creating environment pollution directly or indirectly in many ways. For example, air is polluted by emitting carbon from power stations, various types of industries and vehicles which typically use conventional fuel like oil and gas as energy resources. These activities are causing climate change and global warming, which is very horrendous for us and our planet. We are already suffering some terrible changes of the nature for these activities. If these things continue to happen then our planet will turn into a disaster situation in the future. Moreover, natural resources of our planet like oil, gas, coal etc. are very limited which will be finished in the near future. We can save these natural resources by minimizing our dependency on them and maximizing the use of renewable energy resources like sun, wind, water, biomass, geothermal and tidal wave of the ocean.

So, we are realizing that the overall control of protecting our planet is in our hands. We can easily destroy our nature, again we can protect and reconstruct it. If we can increase awareness among ourselves, we can easily protect our planet and create a Green Planet for our next generations.

Contents

How to size battery for solar panel

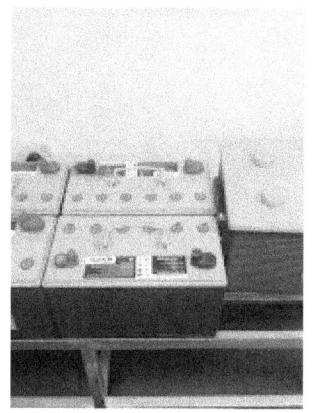

While designing a solar photovoltaic off-grid system, you need to determine required batteries. A little math is necessary to size batteries. An off-grid battery bank should be designed in such a way that the system is not only sufficient enough to supply necessary power during cloudy days but also small enough to be charged by your solar panels. First of all, you need to decide how much storage you want to provide by your battery bank. It is familiar as "days of autonomy" which is based on your expected number of days power provided by the system without receiving an input charge from the solar array. Beside determining days of autonomy, you also need to consider the critical situation and usage pattern of your application. If you decide to install a system for a weekend home, you need to consider a larger battery bank for charging and storing energy all week. But, if you want to add a photovoltaic array to a generator-based system as supplement, then you need to undersized your battery bank slightly as the generator can be operated based on recharging needed.

Rated Capacity of Battery

Basically, the capacity of a battery is specified as Ampere-Hour (AH) along with some specific standard hour reference like twenty or ten hours. For example, you have a battery which is rated at 200 Ampere-Hours and specified as a 20 hour reference. This means the battery is fully charged and will deliver a current of 10 amperes for 20 hours. If the discharge current decreases then the capacity will increase.

Depth of Discharge

The percentage of the rated battery capacity that is withdrawn from the battery is called depth of discharge. The withstand discharge capability of a battery depends on its construction. Batteries are specified by two commonly use terms. They are deep cycle and shallow cycle. Shallow-cycle batteries are comparatively cheap, lighter and have a short lifetime. Stand-alone photovoltaic systems should use deep cycle battery. Deep cycle batteries contain thicker plates and withstand up to 80% daily discharges of their rated capacity. These types of batteries are flooded electrolyte. That is the plates of batteries are covered with the electrolyte. In order to keep the plates fully covered, it is required to monitor the level of fluid and add distilled water added periodically.

Effect of Temperature

Batteries are very temperature sensitive. You cannot expect as much energy out of a cold battery as a warm one. Though, you can get more than rated capacity from a hot battery, operation at hot temperatures will decrease the lifetime of battery. It is recommended to keep your batteries near room temperature. In order to optimize the charging cycle at various temperatures and increase the lifetime of your battery, charge controllers can be purchased with a temperature compensation option.

Lifetime of Battery

It is quite difficult to predict the lifetime of any battery absolutely. Because it depends on various factors like depth of discharge, charging and discharging rate, number of cycles and operating at extreme temperature. It is quite exceptional for a lead-acid battery to last longer than fifteen years in a photovoltaic system. But, usually it can last up to eight years.

Maintenance of Battery

Periodic maintenance is required for batteries. In order to ensure that connections are tight and there is no indication of overcharging, it is necessary to check sealed battery periodically. For flooded batteries, it is required to maintain the plates well above the electrolyte level, check the voltage and specific gravity of the cells for consistent values. If the reading shows a large variation, then it may indicate cell problems. It is necessary to check the specific gravity of the cells by a hydrometer before the onset of winter particularly. The electrolyte in lead-acid batteries may freeze in cold environment. The freezing temperature is a function of a battery state of charge. The electrolyte becomes water and battery may freeze if it is completely discharged.

Sizing the Battery

The recommended battery type for using in solar photovoltaic system is deep cycle battery. Deep cycle battery can discharge to low energy level. It also can be recharged rapidly. In order to operate at night and cloudy days, the battery should be large enough to store sufficient energy. The size of battery is determined as follows:

1. Determine the total Watt-Hours/day consumed by appliances.
2. Divide the total Watt-Hours/day consumed by the battery loss factor (typically, it is 0.85).
3. Divide the answer you got in item 2 by the depth of discharge of battery (typically it is 0.6).
4. Divide the answer you got in item 3 by the Nominal Battery Voltage.
5. Multiply the answer you got in item 4 with days of autonomy (the number of days required for the system to operate by solar panels when there is no power generated) to get the required capacity of the deep-cycle battery in Ampere-Hour (AH).

Battery Capacity (AH) = $\dfrac{\text{(Total Watt-Hours/day consumed by appliances x days of autonomy)}}{\text{(battery loss factor x depth of discharge x nominal battery voltage)}}$

How to choose a solar panel

 Solar panel is one of the important elements in solar photovoltaic system. Therefore, choosing the correct solar panel is crucial. You need to determine the right wattage of a solar panel. The size of the solar panel in watts directly affects the cost which is basically priced in dollars per watt. Watts are related to the output of each module. For example, a 50 watt panel installed under ideal condition will generate 50 watt-hours of electricity each hour and a 100 watt panel will generate 100 watt-hours in each day. So, you can expect to pay double the price for 100 watt panel compare to 50 watt panel.

There are three main stages in solar photovoltaic system. They are:

- Generating power by the solar panel.
- Storing power by the battery.
- Using the power.

In order to determine the wattage correctly, you need to size the panel according to your required power usage. Battery stores power to be used later on. To maintain the constant level power storage in the battery, the solar panel needs to supply the battery same amount of outgoing power.

The cost of solar panel depends on the size (in watts), the physical size, quality of materials, warranty period, brand and certification of the solar panel. The price also depends on how many solar panels you are purchasing as part of full system package. In general, purchasing large amount of solar panels will cost less per unit. But, choosing solar panels based on only the price is not a wise decision. It may not fit the area where you want to install, not having required certifications to qualify for government rebates and lack of solid warranty period.

Besides cost, it is important to consider the manufacturing process and materials used in the solar panel. Since all panel manufactures are not the same, therefore you should consider other factors before the decision of purchasing panels. They are:

- You need to consider the tolerance rate of solar panel. For example, you have purchased a panel mentioned 100 watts in the "nameplate". But, in reality it will be 95 watts only because of quality control issues. Therefore, a positive tolerance rate is crucial. That is, under standard condition, a panel of 100 watts will not only generates 100 watts, but also performs more effectively.
- There is a significant impact on temperature co efficient of solar panel. It is better having less percentage per degree Celsius. The price of solar panel with less percentage of temperature co efficiency is comparatively higher.
- Conversion efficiency of solar panel is also vital. It determines how much power is generated by solar panel during the conversion of light into electrical energy. For example, you have purchased two panels with same price. But, one has higher conversion efficiency than other, then it provides better value for money with correct efficiency.
- In certain climate condition, Potential Induced Degradation (PID) can be caused by stray currents triggered which are the reason for substantial power loss in the panel. Solar panel with little or no PID is considered as good.
- Embodied energy of the solar panel is another crucial factor. Here, panel's initial energy intensive production is compared with the time of pay back producing more energy.
- After installing solar panels, Light Induced Degradation (LID) may occur within few months which decreases the amount of power produced in the module. If there is little or no LID in the panel, then the panel is considered as good.
- Based on your installation application, the best type of solar cell will vary for you. There are three major types of solar cell remain. They are mono -crystalline, poly-crystalline (or multi-crystalline) and amorphous (or thin-film) silicon. High efficiency and good heat tolerance characteristics remain in mono-crystalline silicon. Due to recent development in poly-crystalline panel technology, it has been observed that panels with poly-crystalline silicon are equal to or better than many mono-crystalline in terms of heat tolerance, size and efficiency. Least amount of silicon is used in amorphous (or thin-film) silicon. Typically, thin-film contains least efficient solar cells. At present, some panel manufacturers are producing thin-film module with highest conversion efficiency.

How to start a business in renewable energy

The demand for energy is increasing rapidly. But, the potential reserves of conventional fuels are limited and it will be finished one day. As a result, the initiative for starting a business in renewable energy is increasing significantly. Renewable energy is clean and green alternative energy which includes solar, wind, biomass, hydroelectric power, geothermal and ocean energy. Here, electricity is generated using renewable energy sources like sun, wind, biomass, water and geothermal heat. Starting a renewable energy business or product can be a profitable effort which contributes positively on our environment as well.

First Step

First of all, you need to determine what kind of renewable energy business are you interest to establish. Because, there are various types of business opportunities remains in renewable and green energy fields with different vision, mission, expenses and operational strategy. For example, solar products affiliate sales program for energy conservation is different from solar panel cleaning business.

Second Step

In the second step, you need to check what permits are required. Check all the necessary requirements of your local government authority like city council, county commissioners or the secretary of state in order to determine what kind of business licenses and permits you need to operate your business legally.

Third Step

Increase your know-how about the green and renewable energy industry. In order to succeed in this industry, your knowledge and passion are required. Study the national recycling, environmental state and energy efficiency statistics, product evaluations and consumer demands for renewable energy sources. Search business opportunities wherein training programs, support, seminars and workshops are included.

Fourth Step

Enlist your potential suppliers carefully. It is very important indeed. Some renewable energy companies can't survive long due to inferior product quality and poor business practice. After determining your investment option, initiate a feasibility study of the business or project of companies you are interested to work with. Before investing, check all available ratings and reviews with the Better Business Bureaus.

Fifth Step

In the fifth step, prepare marketing and business plans. Basically, your business plan is a handbook which helps to guide you and your investors through each step for building your company. Prepare product information, operational methods, an executive summary, financial strategy, vision and forecasting. Determine a solid marketing strategy in order to find and reach your target consumers effectively.

Sixth Step

Try to obtain any necessary capital. You may be eligible for grant funding according to your renewable energy business type. Check whether you meet all the requirements to qualify for any small business grant which awards your local government. In this situation, you will need to submit complete application and written proposal along with copies of your marketing and business and plan. If you need additional funding or you are not eligible for renewable energy business grants, then look for loan scope.

Seventh Step

In this step, select your target consumer, execute and implement your marketing and business plan. After securing necessary funding and establishing start up logistic supports, engage your target consumer for generating sales. For getting most potential customers, research local demographics. Communicate with construction companies, consumers, homeowners, landlords, school districts and other organizations so that you can provide them your renewable energy products or services as per their requirement. It will help to develop business relationship mutually.

Eco Friendly Hotels

A hotel which strives to implement the efficient use of energy, water and materials using environmentally friendly techniques while providing quality services is considered as Eco friendly hotel.

Eco friendly hotels concentrate to minimize its impact on the environment. Here, the practices of green living are strictly followed. They are certified green by the appropriate authority of the state where they are located in. Typically, these types of hotels are often located in jungles as Eco lodges. An Eco lodge is sustainable merged with natural, built and social environments.

Now days, Eco friendly hotels endeavor to improve their "Green" credentials locating in less "Natural" location. In order to ensure a safe, non-toxic and energy-efficient accommodation for guests, Eco friendly hotels strictly follow green guidelines. An Eco friendly hotel consists following basic characteristics:

- Entire environment is completely non-smoking.
- Local and sustainable building materials are used during construction of the building.
- Buildings are typically built in harmonic with the natural surroundings.
- Building consumes less than 1/3 of the overall land area.
- Traditional designs are reflected in the building.
- Mattresses, sheets and towels are 100% organic cotton.
- Non-toxic natural cleaning agents and laundry detergents are used in housekeeping.
- Utilizes renewable energy sources like solar and wind energy.
- Entire lighting system is energy efficient.
- Utilizes daylight system.
- Uses organic soaps and shampoos.
- Uses biodegradable nappies.
- Having provision for recycling organic and non-organic wastes.
- Green vehicles are used as on-site transportation.
- No chemicals are used in the food production system and organic foods are available for guests.
- Uses non-disposable dishes.
- A fresh natural air exchange and cooling system is available.
- Employs sustainable waste and water management system.
- Supports biodiversity and plants native trees annually.
- Informs guests, visitors and staffs about significance of healthy ecosystems and describes how to best enjoy the area without causing negative impacts.
- Encourages the development of sustainable economic community.
- Helps and demonstrates the local economy that ecotourism is more sustainable long term way to earn income than destroying habitats for short term gain.

Either conviction or fashion, ecological technologies are very strong trend now a days. They can contribute to protect the environment from negative impacts significantly. As a result, the popularity of Eco friendly hotels are rapidly increasing in the tourism industry. In terms of socioeconomic and environmental context, Eco friendly hotels contribute to develop local economy and help to reduce carbon emissions effectively. Therefore, the demands for establishing a large number of Eco friendly hotels are increasing rapidly.

How can make your home environmentally friendly?

 Making your home environmentally friendly can contribute to protect the environment significantly. You can follow and apply some techniques in order to make your home environmentally friendly. Required products for implementing these techniques are comparatively inexpensive. You can purchase these products at your local home improvement store. Following techniques should be followed and implemented in order to make your home environmentally friendly:

Minimize your electrical usage:

A large portion of a home utility bill is caused by electrical usage. For example, when you plug any household item into an outlet, it is utilizing energy even if the item is not on. For creating a baseline for home electrical usage, an energy audit should be conducted.

How can you conduct an energy audit?
- You need to record when the item is used or requires electricity.
- Make an evaluation when the item requires to be connected to an outlet.
- You should record each and every item which is connected to an electrical outlet.
- You need to shut off any unused surge suppressors and power strips.

During purchasing a large appliance, it is very important to make sure that the appliance is more energy efficient. At present, incandescent bulbs are used in many older homes. Initial costs of these types of bulbs are comparatively low but are highly inefficient. Incandescent bulbs should be replaced with Compact Fluorescent Light (CFL) bulbs. Though the initial cost of CFL bulb is comparatively higher than incandescent bulb, the energy required to power the bulb is significantly less. Light Emitting Diode (LED) is also a good choice for most home lighting systems. Like a spotlight, LEDs emit the light in a specific direction. The lifespan of LEDs are comparatively higher than incandescent and CFL bulbs. LEDs are cool to the touch, so they don't consume much energy at all. In comparison, they are still more expensive than either incandescent or CFL bulbs.

Maximize the efficiency of heating and cooling systems

Typically, heating and cooling system users are the largest electrical users in any home. The most efficient system currently available is geothermal heating and cooling system. Installation of a heat pump, heat ex-changer and distribution system are required for this system. The initial cost of this type of system is comparatively higher than that of the average heating and cooling system, but is still reasonable due to the rapid decrease in electrical utility bills once it is installed. A significant amount of energy will be wasted due to unnecessary cooling and heating if the doors and windows in a home are not sealed properly. You can purchase several products in order to seal your doors and windows. It is very important to inspect all seals in a home annually. Rope Caulk can be utilized to create a seal where windows open and close. Alternatively a Window Insulation Kit can be used to place an air tight plastic covering over the window.

Use Eco paints on your walls

Traditional paints contain damaging volatile organic compounds (VOCs) which can continue being emitted into your home's atmosphere for 5 years after painting. You should use plant-based and water-borne paints. If you can't find plant-based paints, try to find paints that are labeled "VOC-free". Many large paint manufacturers produce paint which is VOC-free.

How do solar cells work?

Solar cell is an electrical device which converts the light energy directly into electricity utilizing photovoltaic effect. It is also defined as the form of photoelectric cell having electrical characteristics like current, voltage and resistance.

There are at least two semiconductor layers in solar cells. One layer contains a positive charge and the other layer contains a negative charge. A typical silicon solar cell consists of a thin wafer having phosphorus doped (N-type) ultra-thin silicon layer on the top of boron-doped (P-type) thicker silicon layer. When these two materials are connected with each other, a junction is formed which is called P-N junction. As a result, an electrical field is created near the top surface of the cell. When these two layers are connected to an external load, the electrons flow through the circuit generates electricity.

Sunlight consists of small particles of solar energy called photons. When sunlight strikes on the surface of solar cell, many of the photons are reflected and absorbed by the solar cell. Electrons are released from the negative layer of semiconductor material, when enough photons are absorbed by this layer of the solar cell. These electrons naturally move into the positive layer and create a voltage differential.

Under open circuit, no load conditions, a typical silicon solar cell generates about 0.5 – 0.6 volt DC (Direct Current). The output of a solar cell depends on its surface area (size) and efficiency. It is also proportional to the light intensity of the sun striking the surface of the cell. For example, under peak sunlight conditions, a typical commercial solar cell with a surface area of 160 square centimeters will approximately generate peak power of 2 watts. If the intensity of the sunlight is 40 percent of peak, then the cell would generate approximately 0.8 watts. In order to increase the output power, cells are combined in a weather tight package which is called a solar module. In order to create the desired voltage and amperage, these modules (from one to several thousand) are then wired up in series and parallel with each other. It is called a solar array.

The semiconductor material silicon which is primarily used for the manufacturing process of solar cells is naturally available. Due to the natural availability of silicon and the practically unlimited resource in the sun, solar cells are very environmentally friendly. Solar cells burn no fuel and have absolutely no moving parts which makes them virtually maintenance free, silent and clean.

Ocean Energy

What is Ocean Energy?

All type of renewable energy which is acquired from the sea is called Ocean Energy. Constant flow of ocean currents contains huge amount of water across the earth's ocean. Technological development contributes to extract energy from ocean currents and convert it into usable power.

Constantly moving ocean waters are affected by water salinity, wind, rotation of the earth, temperature and topography of the ocean floor. Wind and solar heating of surface water near the equator contribute to drive most ocean currents. Meanwhile, salinity and density variations of water column create some currents. Ocean currents are comparatively constant and flow in one direction. Due to the density of water, ocean currents move slowly and contain great amount of energy in comparison to typical wind speed. Water is denser than air. For these characteristics, ocean energy can be captured and converted into usable form of electricity.

Ocean Energy is classified in three types. They are as follows:

- Mechanical energy from the waves that is wave energy.
- Mechanical energy from the tides that is tidal energy.
- Thermal energy from the sun's heat.

How does it work?

Wave energy: This type of energy is produced by converting the energy of ocean waves or swells into other type of energy which is only electricity at present. Various types of technologies are developing and trialing in order to convert the energy of waves into electricity.

Tidal Energy: Tidal movements contribute to generate tidal energy. Both potential energy and kinetic energy exist in tides. Potential energy is related to the vertical oscillations in the sea level. On the other hand, kinetic energy is related to the horizontal motion of the water. Technologies of producing energy from the rise and fall of the tides can be used to harness it.

Ocean Thermal Energy: By converting the temperature difference between surface water and water at depth into useful energy, ocean thermal energy can be generated. Ocean thermal energy conversion (OTEC) is a system of twenty four hours base load electricity generation per day throughout the year. In comparison to other ocean energy sources, OTEC is one of the continuously available renewable energy resources which contribute to base load power supply.

Green Economy

Green Economy is the economy wherein sustainable society exists with zero carbon emissions and a one-planet footprint. Here, naturally restored renewable resources are utilized to acquire energy. A green economy is applicable to people, planet and profits at the micro and macro-economic level of all organizations. Meanwhile, the foundation of "Black" energy economy exists with carbon-intensive fossil fuels such as petroleum and coal. On the other hand, a low-carbon economy is different from a green economy as carbon emissions are still created by it.

Characteristics of Green Economy

Here, the infrastructure is completely powered by renewable energy. It utilizes green energy technologies and also regulates the carbon market in order to build up a sustainable green economy.

Green economy can contribute to balance our ecosystems, biological diversity and forests creating sustainable model of governance, business models and markets wherein sustainable ecosystem services are maintained for the successful adaption to climate change at a local, regional and global level.

Green economy consists of industries like ecosystem services, climate change adaptation services, bio fuels, biomass, carbon capture and storage, carbon markets and renewable energy credits, distributed generation, energy storage, energy recycling, energy conservation, energy efficiency, energy controls and software, green information technology, green design, green materials and construction products, green building, solar energy, wind energy, geothermal energy, hydro power, ocean power, sustainable and organic agriculture, waste and waste water management etc.

Due to high carbon emissions, coal and petroleum are not considered as a green economy. Though nuclear energy comparatively generates very little carbon emissions, it is excluded from a green economy due to its potential impact on environment.

Significance of Green Economy

Green economy significantly contributes in the economic growth of a country. It enhances the socioeconomic growth in a sustainable way indeed. Protecting the environment is one of its significant contributions. Through green economy, green employments can be generated which is an important factor for growing economy. It assists the government and organizations to move forward to implement the challenge of future economy through sustainable green method.

Green Business

What is Green Business?

A green business is a business which has minimal negative impact on environment, community, society and economy. It develops business policies and demonstrates commitment to a healthy and sustainable future. A green business should contribute to enhance the quality of life for its employees and customers. Now days, certification systems have been introduced which strive to standardize these policies. Green Business should meet following requirements:

- Business decisions and policies should be implemented following all the principles of sustainability.
- The business operation should create a significant commitment to environmental principles.
- Supplying products should be significantly green and environmentally friendly.
- The business is traditionally greener.

Strategies for Green Business

Green Business has significant impact on ecological and social performance creating it into a three-part organization model which operates along three axes: people, planet and profit. That is, companies should not only consider profit in measuring a business success but also think about social and environmental factors. So, a green business should consider the following environmental and society strategies:

Energy Management: Energy management is one of the important factors in green business. Sustainable businesses understand that climate change will have a significant impact on our world at all levels, including business success. Green businesses that follow balanced energy management system in planned way, not only save money, but also save themselves against the risk of increasing energy costs and will establish their interest and commitment regarding climate change seriously. Manufacturers should follow energy savings methods for their factory floor and also contribute to produce products that require less energy.

Water preservation: Green businesses realize that supplying fresh water is very limited which requires preservation. During manufacturing and distribution process, the company should try to minimize water consumption of water and also should strive to produce products which consume less water during their lifespan.

Solid waste saving and recycling: Quantity of waste produced during the manufacturing process should be minimized. Recycling and reusing of these wastes is the keystone of any green business. As usual, the finished product should also minimize the waste production.

Preventing Pollution: Green Business practices preventing pollution. Sometimes, during manufacturing processes toxic ingredients which causes air pollution, water pollution and ground pollution. A green business will effort to minimize toxins going into their products, diminish toxins at the end of the process and contribute to create toxin-free products.

An ideal green business should follow above strategies. They should plan and produce products that really help the environment. Whether a green business chooses to manufacture highly-efficient solar panels or engineer a new water purification method, the world requires more products that will not only have less impact but also contribute to develop our planet significantly.

Green City

Green City is the system of creating a green and sustainable city by utilizing and implementing green technologies and policies. It includes renewable energy generation, environmental impact per person, environmentally friendly green transport used by people, recycling programs, constructing green building and reserve green spaces. In order to achieve green status, it is required to improve the quality of the air, minimize the dependency on non-renewable resources, encourage the building of green homes, offices and other structures, reserving more green space, support environmentally friendly methods of transportation and offer recycling programs. Following implementations are necessary for creating a Green City:

- Appropriate urban planning should be made comprehensively.
- Location with green natural beauty makes people feeling a connection to their surroundings.
- Going green not only save the planet but also contributes to generate revenue for a city. Sustainable manufacturing and services can contribute to make money.

Most sustainable cities of the world follow some best practices of green technologies and policies. They are as follows:

- Committed and appropriate plan with regular progress report.
- Building codes should be implemented by green technology appropriately.
- Generating electricity utilizing renewable energy resources like sun, wind, geothermal and biomass etc.
- Implementing tree plantation.
- Encourage people to buy and use environment friendly green products.
- Creative and knowledge based economies should be encouraged.
- Green transportation system for public should be invested and encouraged to minimize the carbon emissions.
- Enhanced density.
- Endeavor and policies to minimize the waste and water consumption.
- Affordable and healthy food should be approached.
- Citizens should be encouraged to involve in the efforts of grass roots.
- Example should be made by City Mayor.
- It is a good thing to create a culture of little competition.

At present, many city mayors are planning and working to get their cities concentrated on the environmental movement. Their goal is to convert their city into a green city. In order to obtain green status, they are endeavoring to improve the air quality, minimize the use of non-renewable resources, encourage to construct environmentally friendly green building, offices and other structures, support environmentally friendly green transportation methods, reserve green space and offer recycling programs.

In order to make our cities green, a lot of green technologies and policies should be implemented appropriately. Cities like San Francisco, Vancouver and Copenhagen etc. are endeavoring to become the world's greenest city. Enhanced awareness and minimized consumption of resources are the ultimate results of a Green City.

Bioelectricity

Bioelectricity is the process of producing electromagnetic energy by living organisms. The bioelectric activity which happens throughout the human body is very necessary to life. Living cells can produce electric, magnetic and electromagnetic fields which enable the action of muscles and the transmission of information in the nerves. This is the concept of quick signaling in nerves. It produces physical processes in muscles or glands. There is some similarity among the muscles, nerves and glands of all organisms. The early development of fairly efficient electrochemical systems is the reason behind it. Scientists are concentrating on the muscles or nerves tissues like the brain, heart, eye, ear, stomach and certain glands, electric organs in some fish and potentials associated with damaged tissues.

Electric movement in living tissue is a cellular concept which depends on the cell membrane. This membrane works as a capacitor wherein energy is stored as electrically charged ions on reverse sides of membranes. The stored energy is available for quick operation. It steadies the membrane system so that it is not activated by small disturbances. Cells capable of electric movement show a resting potential wherein their interiors are negative by about 0.1 volt or less compared with the outside of the cell.

Bioelectric signals are triggered by electrically active tissues like the brain, heart or the muscles. These active tissues can cause some concentration differences in the extra-cellular fluid which includes sodium, potassium and chloride ions. This is why one can measure signals like ECG or EEG from outside the body on the surface of the skin, with the help of electrodes. An interface between the extra cellular fluid and the metal of the wire is constructed by the electrode. The electrode is a sensor which consists of a metal and often a salt-bridge. Here, the local differences of the concentration of charged ions are converted into an electrical signal. The bioelectric signal which is measured from the skin's surface is within the approximate range of 0-2000 µV (2 mV).

More electrical phenomena exist inside our body or on the electrode. Two of them are the DC (Direct Current) offset of the electrode and the 50 or 60 Hz (Hertz) mains interference or main potential. Furthermore, any measurement will display the noise which is produced by the body, the electrode impedance or the amplifier itself. Measuring potential differences between two points on the body can provide very important information regarding the electrical activity happens inside our body. All these noise and signals are dealt by the measurement configuration in such a way that the bioelectrical signals which are measured on the skin's surface are reflected in the output signal positively and cleanly.

Solar Boat

Solar boat is an electrical boat which is powered by solar energy utilizing solar photovoltaic modules, batteries and other necessary electrical accessories. They are quiet, independent and clean engines. Here, batteries store free energy from the sun. The available sunlight is converted into electric power by solar cells which are temporarily stored in batteries. It is used to drive a propeller through an electric motor. Typically, power levels are within a few hundred watts to a few kilowatts. A specific solar boat can run on solar energy depends on its technical design, the amount of photovoltaic cells carried, the solar climate where it is based and its pattern of use.

There are very few grid connections on the seas and inland waters as well as along their banks. People living on inland waterway crafts, houseboats, sailing boats and space stations are dependent on batteries just like the owners of electrically propelled boats. But batteries sometimes discharge and need to be recharged. In this scenario, solar photovoltaic system may be one of the significant solutions. On a ship, solar photovoltaic modules can charge the batteries on the spot effectively utilizing free energy from the sun without polluting environment. Electric motors are used in submarines and cruise ships for the momentum. Large passenger liners, aiming to provide maximum comfort of traveling, do not however store their energy in batteries but produce it by using many small diesel generators. They create noise and are not environmentally friendly technology. On the other hand, solar photovoltaic system generates electric power without any noise and environment pollution. This technology is reliable and durable.

Solar photovoltaic modules can be installed into the boat in reasonable areas in the deck, cabin roof or as canopies. Some solar photovoltaic modules, or photovoltaic arrays, can be flexible enough to fit to surfaces which is quite curved and can be ordered in unusual sizes and shapes. However, the weightier, inflexible monocrystalline solar photovoltaic modules are more efficient in terms of energy output per square meter. After installing solar photovoltaic modules on the boat or ship, the efficiency of solar cell decreases, because they must be covered in layers of plastic and transparent composites in order to make them protected to the destructive effects of seawater and variable temperatures. Therefore, it is important and advantageous to tilt the photovoltaic arrays.

Bioplastics

Bioplastics are the bio based plastics produced from renewable resources like corn starch, pea starch, vegetable fats and oils. On the other hand, conventional plastics or fossil fuel plastics are produced from petroleum. Conventional plastics create more greenhouse gas which is very dangerous for our environment.

Petroleum is very limited resource in the earth. It becomes expensive day by day. One day, this resource will be finished. During burning petroleum products like plastics, carbon is emitted and it causes the climate change. Conventional plastics are harmful by-products and chemicals. In this situation, bioplastics can offer an alternative solution. Sugarcane ethanol is a significant substitute of petroleum in the production of plastic. These bioplastics contain same physical and chemical properties like conventional plastic. They also maintain full recycling capabilities.

Applications of Bioplastics

Bioplastics are used in packaging items. They are also used in the items like bowls, pots, cutlery, crockery, straws, trays, bags, bottles for soft drinks and dairy products as well as containers for eggs, fruits and vegetables. Other applications include fuel line and plastic pipe applications, carpet fibers, car interiors and casings of cell phones.

Benefits of Bioplastics

- Bioplastics producing component like sugarcane polyethylene substitutes 25 percent or more of the petroleum which is very significant for our environment indeed.
- In order to produce bioplastics, less than half the energy is required in comparison to the production of conventional plastics. So, using the same amount of energy, we can produce twice the amount of biodegradable packaging and bags.
- Bioplastics are safe. There is no chemical or toxin in bioplastics. It can be composted locally into a soil amendment.
- Maximizing the use of bioplastics contributes to minimize the dependency on fossil fuel which is very significant.

Prospects of Bioplastics

Bioplastics are gradually appearing as a promising substitute of conventional plastics. Now, they are playing a significant role for a modern and forward looking plastics industry in the context of environment as well as economic development. Bioplastics are rapidly growing industry. In order to meet the highest standard, bioplastics technology is upgrading rapidly. Still, there is the scope to develop this technology for the betterment of plastic industry. Instead of utilizing edible resources, effective research development is taking place utilizing inedible organic waste resources as well as giving the products more solidity and functionality. As a result, application areas of bioplastics are expanding very rapidly with good prospects.

Solar Water Disinfection

We know that water is an important element in our life. Our lives are survived by water. Indeed, our body contains huge amount of water. Purified water can save our life. On the other hand, germ infected water can destroy our life. So, we need to drink water which is disinfected. We can make our water disinfected by utilizing various systems. Solar Water Disinfection is one of the significant systems to make our water disinfected.

Solar Water Disinfection is a system which utilizes solar energy to remove biological agents like bacteria, protozoa, viruses and worms from water and makes it disinfection. There are three types of method to disinfect water by solar energy. They are as below:

Method-1: Utilizing the effects of electric current produced by solar photovoltaic module.
Method-2: Solar thermal water disinfection.
Method-3: Disinfecting water by utilizing solar ultraviolet.

In the first method, solar photovoltaic modules are used to produce electric current which provides electrolytic processes in order to disinfect water.

In the second method, heat energy from sun is utilized to heat water for a small period of time. Here, solar heat collectors use reflectors and changing levels of insulation. Some solar thermal water disinfection processes are batch based and others are through flow based. During full sun shine, through flow based process operates almost continuously.

In the third method, sunlight and plastic PET bottles are used to disinfect water. It is a free and effective method for decentralized water treatment. This method is known as Solar Ultraviolet (UV) Water Disinfection and familiar as SODIS. Typically, it is used in household level.

In the SODIS method, contaminated water is filled in a transparent PET bottle and exposed to the sun for 6 hours. During this period, UV radiation of the sun kills diarrhea causing bacteria pathogens. It has been observed by researchers that UV-A solar radiation can easily destroy microorganisms situated in water. If you would like to observe whether the water is clear enough for the SODIS method to function appropriately, then you can precede the newspaper test. In this test, filled bottles are placed upright on top of the newspaper headline. Now, it is required to look down through the bottle opening. If you observe that letters of the headline of newspaper are readable, then you can use the water for the SODIS method. If the letters are not readable, then you must treat the water before applying to the SODIS method. This method contributes to prevent diarrhea and save lives of people.

Biomass

Biomass is the biological component which is obtained from living organisms of nature. It is also referred as plant based material when energy is produced from biomass. At the same time, biomass is also applicable for both animal and vegetable resultant components.

Biomass is a renewable energy source and it can re-grow quite quickly. Sun's energy is captured by plants' chlorophyll through the photosynthesis process. Here, carbohydrates are created by converting carbon dioxide from the air and water from the ground. They are complex combination of carbon, hydrogen and oxygen.

Basically, plant materials are either broken down by micro-organisms or burned. After burning, carbohydrates are transformed into carbon dioxide and water and release the energy harnessed from the sun. This process is known as the carbon cycle in the earth. In this technique, biomass works like natural battery in order to store solar energy. Appropriate utilization of biomass can enhance the longevity of this natural battery as well as provide low-carbon energy sources. Sustainable low-carbon biomass is great source of new renewable energy which is environmentally friendly indeed.

Types of Beneficial Biomass

There are following types of beneficial biomass in this planet:

- Wood and forest residues which are harvested in sustainable way.
- Crop residues like wheat straw.
- Energy crops which don't compete with food crops for land.
- Clean industrial and municipal wastes.

Typically, beneficial biomass sources help to maintain and increase the stocks of carbon stored in plants or soil. It also moves emissions from fossil fuels, like coal, natural gas and oil. Most sustainable and effective biomass resources may be varied from region to region. It also depends on the types of final application you wish to use. For example, you can utilize biomass as bio power, bio products, heat and biofuel.

How to convert Biomass into Bio power

Generating heat energy by burning biomass is one of the most familiar methods since long time. This biomass fired heat can produce steam power which can be utilized to produce electricity. Burning biomass in conventional boilers provides environmental and air-quality benefits over burning fossil fuels. Recent researches by researchers indicate that biomass can be converted more cleanly into liquid fuels. Producing combustible gases from gaseous process is one of these examples. It decreases different kinds of emissions from biomass combustion.

Benefits of Biomass

- Typically, biomass offers environmental, economic and energy security benefits.
- Biomass energy decreases air pollution and net carbon emissions.
- It contributes to protect the quality of soil, maintain wildlife habitat and avoid erosion.
- Biomass helps to minimize the dependency on importing fossil fuels which reduces our expenses.
- Rapid growth of biomass energy technology helps farmers and forest owners to obtain valuable new markets for their new energy crops, forest and crops

residues.

- It contributes to minimize the global warming.

Hygroelectricity

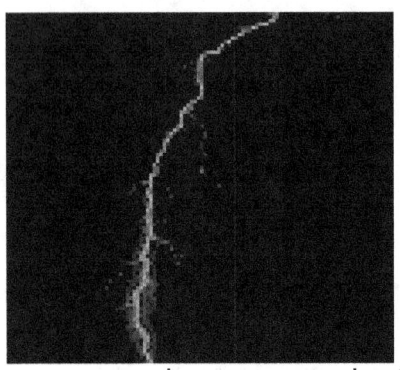

For long time, It has been a great mystery how electricity appears in the atmosphere. But, recent research says that atmospheric water vapor can turn into electrical charge. They are calling it hygroelectricity which means humid electricity.

Now, it is realized that hygroelectricity is one kind of static electricity which appears on water droplets and can be moved from droplets to small dust particles. It is a common phenomenon in the earth's atmosphere. Hygroelectric charge exists in thunderstorms volcanic eruptions and some dust storms which creates lightning. Once, scientists believed that water droplets in the atmosphere were electrically neutral and remain same even after coming into contact with the electrical charges on dust particles and droplets of other liquids. But, recent research says that an electrical charge can be picked up by water in the atmosphere indeed.

Scientists had been charmed by the idea of harnessing the power of thunderstorms for long time. After laboratory experiments, scientists observed the simulated water's contact with dust particles in the air. In their experiment, tiny particles of silica and aluminum phosphate were used. It has been observed that during high humidity, silica gains more negatively charged and aluminum phosphate gains more positively charged. A high level of water vapor in the air creates high humidity. During steamy summer days, this vapor can condense and appears as fog on windows of air-conditioned building and cars.

Scientists and researchers are hoping that just like the solar cells which collect the sunlight to produce electricity, it will also be possible to develop collectors in order to capture hygroelectricity and distribute it to homes and businesses. Just like solar cells work effectively in sunny areas of the world, hygroelectrical panels would work more effectively in areas having high humidity. They propose that hygroelectrical panels can be installed on top of buildings in the area where thunderstorms are most frequent. These panels could drain electricity out of the air and prohibit the building of electrical charge which is released in lightning. Now, scientists and researchers are comprehensively experimenting as well as researching the greatest potential for use in capturing atmospheric electricity and preventing lightning strikes. So, we can say that hygroelectricity contains a good prospect as an alternative energy resource.

BIPV

Full abbreviation of BIPV is Building Integrated Photovoltaic. This is a technology which integrates photovoltaic modules into the roof or exterior of a building. By concurrently serving as building, a BIPV system serves as both envelope material and electrical power generator. It contributes to save materials and electricity costs as well as minimize the dependency on fossil fuels. It reduces carbon emission and enhances the architectural value of a building.

Most of BIPV systems are connected with available grid power. We can also use them as stand-alone, off-grid systems. A grid-tied BIPV system contains mutual utility policy which creates the storage system free indeed. It is highly efficient and contains unlimited capacity. Both the building owner and grid power company are benefited from grid-tied BIPV. It's characteristics of onsite electricity production helps the building owner to reduce energy costs. On the other hand, produced solar electricity can be exported to grid Power Company during peak demand. A complete BIPV system has following components:

- Typically, PV modules of BIPV system are thin-film, transparent and semi-transparent.
- In a stand-alone system of BIPV, a charge controller is used to control the incoming and outgoing power of the battery storage bank.
- An inverter which converters DC output of PV modules into AC output.
- Proper hardware for mounting and structure, wiring and safety disconnects.
- A power storage system.
- Backup power supplies.

BIPV system has following forms:

- At present, thin film solar cells integrated to a flexible polymer roofing membrane are widely used in BIPV system.
- Modules having multiple roof tiles shape are also available.
- Solar shingles are protective against ultraviolet rays and deterioration of water. It removes condensation because the dew point is kept above the roofing membrane. In this way, it protects insulation and membranes and enhances the normal life of roof of a BIPV system.

- Traditional building materials are rapidly replaced by solar facades. It generates income through Feed-In Tariffs as well as saves all costs of traditional materials.
- Installing facades on existing buildings will create a new appearance of old building. These modules are mounted on the facade of the building, over the existing structure, which can increase the appeal of the building and its resale value.
- Semitransparent modules are great substitute of architectural materials constructed with glass, skylights and windows.

Benefits of BIPV System

- BIPV system is an environment friendly technology which converts solar radiation into electrical power without polluting environment.
- Cost of cladding with PV is comparatively lower than conventional cladding materials. So, building owners are economically benefited from BIPV System.
- Thermal insulation effect of PV modules minimizes the heat loss of building.
- PV modules attenuate electromagnetic waves. As a result, PV modules can help shield areas particularly susceptible to electromagnetic interference, for example in hospitals.
- Electromagnetic waves are attenuated by PV modules. It helps protective areas which is vulnerable to electromagnetic interference. For example, we can say hospital. Multilayer PV modules of BIPV System are protective against noise.
- Solar cells of BIPV system are embedded in laminated glass which is very protective against burglary. By producing power by BIPV system, building owners can sell electricity to grid Power Company and earn money. So, BIPV is a source of additional income.
- As it is an environment friendly technology and reduce carbon footprint, building owners can be benefited by government incentives.

Eco Friendly Printer

Every day, we use printers in our home and offices. Now days, printers become an important device for documentation. Typically, printers consume huge amount of power and also printers have significant impact on our environment. So, we need to make our printers more environmentally friendly or Eco friendly. Listed below are some techniques to make our printers Eco friendly.

Reusing and recycling printer cartridges

It is obvious that most of expense in a large office or organization is printer cartridges. Huge amount of printing can quickly increase printing cost per year. So, it should be minimize in environmental friendly manner. You can half the price of retail cartridges of printers by using re -manufactured ink and toner cartridges. It's life cycle is also long. Used cartridges can be recycled, refurbished and then refilled with new printer ink. This is called re manufacturing process. It will not only reduce your office expenses, but also reduce your carbon footprint.

Limiting the use of paper

Limiting the use of paper is a good choice to go green. You can only print documents which are mostly required to print indeed. Most documents can easily read and review in your computer. It will help to save papers as well as reduce your printing cost in environment friendly manner.

Saving Energy

After finishing your work, shut down your computer. If you have a meeting which will continue for more than two hours, then you should set your computer to idle mode in order to save energy. Many office printers also have a sleep mode.

Regulating additional environmental waste using re-manufactured printer cartridges

Uses of printers are increasing day by day. As a result, numbers of used printer cartridges are also increased every year. Every year, lots of new printer cartridges are sold out. But, very few people refill their cartridges. Huge amount of printer cartridges are thrown in the trash every year. It takes many years to decompose. These disposed cartridges have negative impact on the environment and our health. It causes the redness of skin, lung issues and even cancer.

Recycling printer cartridges can save huge amount of oil every year. It also decreases the existence of metal and plastic in landfills. You can reduce your carbon footprint of your printer cartridges by recycling your current empty printer cartridges. Increase awareness among other peoples to protect the environment through recycling. Always purchase re-manufactured printer cartridges online. Re-manufactured cartridges are recycled printer cartridges which is cleaned, emptied and refilled. You can get same quality printouts which you get from your existing printer. Most significantly, re-manufactured cartridges are most cost effective than retail cartridges. They also provide significant contribution to our environment by reducing huge amount of harmful toxins released by landfills worldwide.

Vegetable based toner

Research for Eco friendly printings are continuing day by day. Researchers are now introducing vegetable based toner which becomes well suited with most laser printers. Soy oils are used in "Soy ink" cartridges instead of petroleum. Soy contains very low levels of vapor pressure or Volatile Organic Compounds (VOC) and consumes very small amount of power to grow. It is comparatively more Eco friendly than conventional petroleum-based printer ink. Moreover, paper printed with soy ink is comparatively easier to recycle. Soy ink can be removed easily than traditional ink.

Laser printers are comparatively economical to maintain and they also produce high-quality printouts. As a result, laser printers are broadly used in business organizations. At present, many offices are using soy-based toner. If all laser printer users depend on soy-based toner for their printing needs it will save huge amount of petroleum as well as decrease the emission of greenhouse gas worldwide which is very significant for our planet.

Latex Printers

There is no hazardous material in latex printer ink. It is odorless and Volatile Organic Compounds (VOCs) are very much less in comparison to conventional ink. Water-based latex inks are long lasting and protective against water, smudges and scrapes. This Eco friendly ink is fully dry. It does not take huge time to dry like solvent ink.

Eco-Solvent Ink

This type of ink is almost odor free and also Eco friendly. They are protective against water for up to three years outdoors. It does not contain any Volatile Organic Compounds (VOC) or harmful components. They are available in varieties types of colors. This type of ink does not require any ventilation which is advantageous in terms of our health and environment.

Green Furnace

The furnace which uses renewable sources like natural fuel for their operation is called green furnace. Typically, a green furnace uses olive pits, nut shells, wood, wood pellets from forest residues or surplus industries etc. Green furnace is fed by bio fuel which is more cost effective than conventional fuels like diesel, propane etc. Moreover, due to stability of the price of bio fuel, uses of green furnaces are rapidly increasing in order to generate heat in any home heating system.

Biodiesel is profitable due to its high calorific value per unit weight. It appears as a form of renewable energy and contributes to protect our environment and energy sources of this planet. If we compare it with fossil fuels, we will observe that one kilogram of pellets contains half the calorific value than a liter of diesel. That means, in order to produce same amount of energy like one liter of diesel, two kilos of pellet or olive stones are required.

Types of Green Furnace

Pellet: In these furnaces, components like olive pits or pellets are used. They are absorbed into the boiler by means of screw or suction. Generally, they are used for domestic purposes in order to generate medium power.

Wood stoves: Here, wood logs are used for combustion. Typically, wood furnaces are used at home. So, it is used for domestic purposes.

Poli-combustible: Here, crushed components are needed in order to be combusted. Their size is comparatively large. They are typically used in industrial activities.

Operation of Green Furnace

A green furnace works on the same principal like a gas boiler. Here, fuel pellets are burnt in the burner and produce a horizontal flame which enters into the boiler, as like as diesel systems. During this combustion process of natural fuel, heat is produced. This heat is transmitted to the attached water circuit in the heat exchanger integrated with the boiler. The Produced hot water is used for domestic use, heating pools etc. Heating can be proceeded by any conventional water system like fan convector, radiators or under floor heating system.

During this biomass combustion, a container is required in order to reserve bio fuel situated near the boiler. From the same, a suction or screw feeder, leads to the boiler, where combustion process happens. In order to insert into the boiler appropriately, the pellet type fuel must be reserved at an angle of approximately 45 degree.

Some ashes are produced after burning biomass. These ashes are collected automatically in an ashtray which should be emptied four times a year. For improved operation of the green furnace, a battery can be installed in order to store heat as like as a solar photovoltaic system.

Benefits of installing a Green Furnace

- **Environmentally friendly furnace:** There is no emission of carbon dioxide gas. So, it is an environmentally friendly furnace.
- **Cost effective energy:** Price of biomass is comparatively lower than conventional fuels. It's price does not depend on international markets like fossil fuels.
- **Safe energy:** Unlike the gas, biomass cannot explode.

Hydroelectric Power

Hydroelectric power is the technology of converting the kinetic energy of flowing or falling water into electrical energy.

A dam is constructed where there is a big river or a natural lake in a valley. Water is reserved in the reservoir of the dam. There is the water intake near the bottom of the dam. Dam creates pressure so that water can produce more electrical power. Gravitational potential energy stored in the water and this energy is utilized to rotate turbines of generators in order to generate power. These turbines are constructed within tunnels in the dam wall. With great pressure, water flows inside these tunnels and helps to rotate these turbines. Water can provide huge pressure due to the great height at which is kept in the dam. If the difference of height between the water levels and the water where it flows is very large, then there is the possibility to gain more power out of the water for its greater potential energy. This difference in height of the water is called the head.

Basically, a generator is a device which can convert mechanical energy into electrical energy. Generators work on the principal of electromagnetic induction discovered by Michael Faraday in 19th century. Faraday discovered that flow of electric charges could be induced by rotating an electrical conductor like a wire which holds electric charges in a magnetic field. This rotation creates a voltage difference between two ends of electrical conductor. It causes the electric charges to flow and as a result electric current is generated.

There are two main parts exist in the generator. They are the rotor and the stator. The rotor is the part which can rotate. It contains magnetic field. On the other hand, the stator is the stationary part. Coils or insulated windings are located in slots near an air gap in the stator core. When the rotor is rotated, alternating current (AC) is induced in the stator. Alternating characteristics of the current is produced by changing the polarity of rotor. The generated voltage is proportional to the strength of magnetic field, number of coils and number of windings in each coil and rotating speed of rotor.

Comparison with other power generation methods

Hydroelectricity removes emissions of the fuel gas from combustion of fossil fuel. It significantly contributes to eliminate harmful pollutant like carbon monoxide, nitric oxide, sulphur dioxide, dust and mercury in the coal. In comparison to nuclear power, hydroelectric power does not generate any nuclear waste. Here, no danger of uranium mining or nuclear leaks. Hydroelectricity also avoids coal mining hazards. Hydroelectric power stations have more predictable load factor than wind farms.

Advantages of hydroelectric power

- There is no carbon emission or air pollution during power production by hydroelectric power plant. It is a renewable energy source and environmentally friendly technology.
- Dams can contribute to generate power for many years. After constructing dams, energy is virtually free indeed. It is cost effective.
- Maintenance of dams are not very expensive.
- The sluice gates can be shut when electricity is not necessarily required. It will stop electricity generation. This water can be saved for use another time when there is a peak demand of electricity.
- Stored water can be utilized during peak demands. Water of the lake can be utilized for irrigation purposes.
- Lake which is created behind dams can be utilized for recreational or entertainment activities like water sports and leisure. Sometimes, tourists can observe the natural beauty of large dams.

Green Data Center

A data center is the house of integrated system of computers, telecommunications and storage devices wherein data is stored. Typically, a data center consists of backup power supplies, surplus data communications connections, air conditioning system, various security devices, fire and smoke detecting devices. Big data centers can consume huge amount of power. It is not only costly but also it can pollute air by exhausting smoke of fuel which is burnt in diesel generator to generate power for the data center.

So, we can realize the environmental impact of a data center. We can make data centers more cost effective and minimize environmental hazard by introducing some green techniques. These green techniques are as below:

Data center infrastructure management

Data Center Infrastructure Management (DCIM) can help taking following decisions:

- Monitoring, locating and managing all physical assets of entire infrastructure with a merge view.
- Automating the commissioning of new equipment.
- Minimizing power consumption, energy costs and carbon footprint.
- Automating the capacity planning with unparalleled predicting capabilities.
- Arranging IT to the requirements of the business and upholding the arrangement.

Natural air cooling system:

Natural or free air cooling is the process of exploiting outside air to cool data center facilities instead of using huge power consuming air conditioning units. Now a days, data center service providers are creating racks and servers which is capable to run in the temperatures up to 27°C. In this scenario, natural air cooling will be a significant choice. In order to get natural cool air, it is necessary to keep the windows of data center just open. Dust particles can damage server equipment of data centers. In order to catch these dust particles, filters are required in data centers. Filtered air should meet the required level of humidity. Natural air cooling system is very effective for data centers in terms of environmental and economic prospects. As a result, companies are gradually applying this technique in their data centers.

Combined virtual data center

Virtual software enhances the efficiency of existing servers which helps to decrease the power consumption indeed. This approach can help enterprises to rationalize IT resources and exploit the unused processing power of high power storage devices and servers. Combining virtual software, specialized servers and high bandwidth network connection can reduce capital costs of data center and also increase energy efficiency.

Energy efficient servers

Using energy efficient servers in the data center is very significant. Because, they can help to reduce direct electricity costs of IT as well as operating and capital costs of cooling facilities.

Utilizing renewable energy

Large business organizations can utilize renewable energy like solar and wind to power their data centers. It can assist managers to meet environmental requirements.

Cloud computing

Cloud computing can contribute in your efforts of creating green IT. Because, a cloud computing provides maximum exploitation of your CPU (Central Processing Unit). It is the sharing of spare resources between two users or organizations as per their requirement.

Modular data centers

These data centers are portable. It is designed for rapid placement, high density and energy efficiency. These are ready made data center in a box which can be scrambled very swiftly.

Data center industries are now concentrating to create their data centers green indeed. Researchers have established a technique to enclose electronics in a fluid to remain circuits effective even at higher temperatures. High efficient coolers can reduce its Carbon footprint. In order to confirm that only necessary modules are in service, it monitors requirements of power of loads. Other modules stay in a low power standby mode. It contributes to save huge amount of money of data centers.

Fiber Optic Communication

The technology of transmitting data and information from one place to another by converting light signal into electrical signal through the media of an optical fiber is called Fiber Optic Communication. In order to transmit data, glass (or plastic) threads (fibers) are used. A fiber optic cable contains a bundle of glass threads. Each thread is capable to transmit messages modulated onto light waves.

Typically, fiber optics is a very "Green" technology in comparison to other semiconductor electronics. It dissipates very little amount of energy than copper based cables. It also saves huge materials and reduces costs. One single strand of glass can carry equal amount of data carried by thousands of copper cables. It can do over a longer distance without using electronic equipment to reproduce it.

Core components of fiber optic communication system

Fiber optic communication system contains following components:

- Semiconductor lasers, fiber lasers and optical modulators based Optical transmitters. Photo diodes based optical receivers.
- Optical filters and couplers. Dispersion compensating modules.
- Optical fibers having improved properties regarding losses, guiding properties, dispersion and non-linearity. Semiconductor and fiber amplifiers to maintain sufficient signal powers over huge distances of fibers or as preamplifier before signal detection.
- Multiplexers and optical switches. For example, optical add/drop multiplexers (OADMs) allow wavelength channels to be added or dropped in a wavelength division multiplexing (WDM) system.
- Optical switches are electrically controlled.
- Signal regeneration device (electronic or optical re generators) and clock recovery. Signal processing and monitoring electronics.
- Computers and software to regulate the operation of the system.

Advantages of fiber optic

Fiber optic has following advantages over traditional communications lines:

- The bandwidth of fiber optic cables is higher than metal cables. As a result, they can carry huge amount of data and information.
- In comparison to electrical cables, the weights of fiber-optic cables are light.
- Fiber optic cables are less vulnerable than metal cables to interference.
- It is capable to achieve huge transmission rate. So, the cost per transported bit can be absolutely low. Data is transmitted in digital system which is the natural form of computer rather than analog system.

Prospects of fiber optic communication:

Fiber optic communication is the most familiar technology for local area network (LAN). Already, this communication system has been used broadly within metropolitan areas (metro fiber links). Furthermore, telephone companies are gradually substituting conventional telephone lines with fiber optic cables. In future, almost all communication system will serve fiber optics. So, we can say that fiber optic communication system carries good prospects.

Photosynthesis

Photosynthesis is a process of producing glucose by plant using the energy from sunlight, carbon dioxide and water. Here, light energy is converted into chemical energy. During photosynthesis process oxygen is released which absolutely need to breathe and stay alive. Typically, using chlorophyll, photosynthesis process is happened in the chloroplasts. Chlorophyll is the green pigment of photosynthesis. In the early evolutionary history of life, first photosynthetic organisms were developed wherein hydrogen or hydrogen sulfide was the source of electrons instead of water. Globally, photosynthesis can capture huge amount of energy. It is approximately six times higher than our existing power consumption. Photosynthetic organisms can also convert huge amount of carbon into biomass per year. Basically, photosynthesis happens in plant leaves. The upper and lower epidermis, the mesophyll, the vascular bundles and the stomates are parts of a typical leaf. Epidermis is the outer layer of cells surrounding the body of an organism like a plant leaf or our skin. No chloroplast exists in the upper and lower epidermal cells, thus no photosynthesis create there. Primarily, they provide protection for the rest of the leaf. In the lower epidermis, stomates holes are created in order to exchange air. Here, carbon dioxide enters and oxygen exits. Vascular bundles of leaf transport required water and nutrients around the plant. The mesophyll cells contain chloroplasts wherein photosynthesis creates. Chloroplasts are photosynthetic organelle in plant.

Parts of a chloroplast include outer and inner membranes, inter membrane space, stroma and thylakoids stacked in grain. Stroma is the fluid within chloroplasts. Thylakoid is one of the sacs within a chloroplast. Membranes of the thylakoid contain chlorophyll. Absorbing red and blue light, it appears to us as green color. It makes red and blue colors visually unavailable to our eyes. Green light is not absorbed in chlorophyll. So, finally it appears as green color to our eyes. However, it is the energy from the absorbed red and blue light. It is used to do photosynthesis. As plants cannot absorb green light, it cannot be used to do photosynthesis.

In the thylakoid membrane, light reaction occurs and converts light energy into chemical energy. In the light reaction, chlorophyll and few other pigments like beta -carotene are associated. Each of these distinguished color pigments can absorb quite distinguished color of light and transfers its energy into the central chlorophyll molecule in order to do photosynthesis.

Energy collected through the light reaction is stored by forming a chemical called ATP (adenosine triphosphate). ATP is a compound wherein cells store energy. It contains nucleotide adenine bonded to ribosomes and three phosphate groups. Dark reaction happens in the stroma within the chloroplast and converts carbon dioxide into glucose.

Solar Water Heater

Solar water heater is the technology of warming water utilizing heat energy of sun. During unavailability of solar energy, a conventional boiler can be used in order to make the water hotter.

Benefits of Solar Water Heater

- **Whole over the year hot year:** Typically, this technology can serve whole year, though you will require heating the water further with a boiler during the winter season.
- **Reducing your cost:** As sunlight is free, so just after paying initial cost for installation, your hot water costs will be decreased.
- **Decreasing carbon footprint:** It is the technology of green, renewable heating system and can minimize your carbon dioxide emissions.

How does Solar Water Heater work?

In the technology of solar water heater, solar thermal collectors are used. These collectors can absorb sunlight and collect heat. Solar radiation is captured by collectors. In the form of electromagnetic radiation, we get energy from solar radiation. These collectors are fitted on your rooftop. They collect heat energy from sun and utilize this energy in order to heat up water. It is stored in a hot water cylinder. In order to achieve your desired temperature, a boiler can be used as a backup for heating the water further.

Solar water heating panels are of two types:

- Evacuated tubes.
- Flat plate collectors are fixed on the roof tiles.

In order to provide some contribution to heating your home, larger solar panels can be used. The system could provide most of your hot water in the summer, but it is comparatively less during colder weather.

Maintenance

Typically, maintenance cost of solar water heater is comparatively very low indeed. Most solar water heater can provide a warranty up to ten years. Here, very little maintenance is required. After installing the system, your installer should leave written details of any maintenance checks that you can carry out from time to time, confirming that everything is working appropriately.

Most importantly, you may check yourself possible leaks on regular basis. There will have a strong smell, if any leak of anti-freeze occurs. You should contact your installer, If you observe this situation. Basically, you need to monitor your system to check whether it is functioning as per design. If hot water is not produced or the solar pipework is cold (when the pump is running) during warm, sunny days then you need to contact your installer again. Some installation companies offer an annual service check as per their service warranty.

Your system should be checked thoroughly every 3-6 years by an accredited installer. After this period of time, the anti-freeze which is used to protect your system in the winter season will need to be replaced as it collapses over time and reduce your system's performance. Anti-freeze can survive better if the system is utilized all over the year and not left unexploited during the warmest weeks of the year. In a better maintained system, pumps can survive over ten years.

Green Cell Phone

Now days, cell phones are obviously playing a significant role in our life. At present, we can instantly communicate with each other by using cell phone. But, there is an inherent impact on our environment during using cell phones. When we talk or send text messages with each other by using cell phones, we are creating some negative impacts on our environment directly or indirectly. During manufacturing and assembling cell phones, some toxic chemicals are released into the air and water. When we throw out our cell phone, heavy metals leach into the soil of landfills which is very harmful for our environment.

Different cell phone manufacturing companies are now producing "Green" cell phones using materials like bamboo, recyclable plastic water bottles and corn. One green material is not better than another as long as it is recyclable.

During building and assembling parts of cellular phone like circuit boards and electronic cables, elements like bromine and chlorine releases into our environment which creates a hazard to inhabitants living in adjoining areas.

Some elements like Bro-mated Flame Retardants (BFR) of cell phones which cannot disrupt in the air causes abnormal brain development in human being and animals. Gaseous components of mercury and lead pollute air during manufacturing cell phones. Elements like PVC or vinyl infuses the air during both manufacturing and disposal of handsets which can be hazardous. In order to produce electronic components, rare earth metals should be excavated in remote locations.

Huge amount of electronic waste, or e-waste are gathering worldwide yearly. Most of these wastes are huge amount of hazardous waste. In order to decrease toxic effects and creates "Green" grades, a cell phone should contain very little amount of elements like PVC and BFR. Some mobile phones are made of bio-plastic which is free from both PVC and BFR elements. Recycling materials like recyclable plastic bottles, cans may also be used during manufacturing cell phones. It can save significant amount of energy than cell phones made of ordinary plastic.

Conventional cell phones can cause various health problems for us. On the other hand, green cell phones are manufactured by environmentally friendly technology. These cell phones help to reduce carbon emissions from air. They are also friendly enough in terms of our health concern indeed. So, we can realize that cell phone manufacturing companies can contribute significantly by manufacturing environmentally friendly green cell phones for us.

Solar Attic Fan

Solar Attic Fan is a simple and environmentally friendly fan which is installed on the top of the roof or attic. It can create cool healthier environment in your home as well as save your money. It is totally powered by utilizing free energy of the sun. This is why it is called Solar Attic Fan. This smooth and efficient attic vent is solid, quiet and powerful. Typically, a solar attic fan can improve attic air circulation which contributes to develop a healthy and energy efficient home. Solar attic fans pull hot air and moisture out of your attic. During months of summer season, it keeps your attic cooler and contributes to decrease the load on your HVAC system. On the other hand, during months of winter season, it regulates the increase of moisture and ice blocking. A Solar Attic Fan has significant impact on our environment.

Applications of Solar Attic Fan

Typically, solar attic fans are used in residential homes, industrial buildings and commercial offices. Roof mounted solar attic fans are familiar in residential homes. Gable mounted solar attic fan is attached to your existing gable outlet of your home and is powered by a remote photovoltaic module installed on the roof. Garage outlet has an exhaust system to the solar attic fan which can pull hot air out of your garage easily. Curb mounted solar attic fans are perfect for commercial and industrial building. It attaches to a per-built curb and can exhaust up to 280 square meter per unit.

How does a Solar Attic Fan works?

Roof or attic ventilation is an important feature to maintain your home's structure and save energy. Inappropriate attic ventilation develops moisture which increases the growth of fungus and also causes the deterioration of wood. Appropriate attic ventilation can resist the growth of moisture by creating continuous air circulation. Typically, in an ideal ventilation system, air should enter from roof space and escape outside at roof's ridge.

Sun beats down on the roof surface and heats up the steady air inside the attic. A passive ventilation system is required for homes as per building codes. But, passive ventilation does not provide required pressure to create the force of air through the roof space and outside. This is why, a motorized fan is necessary there. An attic fan installed at the ridge of room can pull air in from outside, force it to move through the entire roof space and escape through the attic vent. During months of summer, a solar attic fan can decrease temperature of roof area up to 40°F. No additional electrical wiring and structural changes are needed in a solar attic fan.

Advantages of Solar Attic Fan

A solar attic fan helps to decrease the interior temperature of a home. During day time, the temperature of a roof can raise up to 160°F or more which causes increasing the interior temperature of your home. An attic fan pulls the hot air from the attic and moves it outside. A solar attic fan can create the inside of a home cooler and more comfortable by venting hot air in the attic outside.

By decreasing the temperature of a home, it helps air conditioning and home cooling systems to work more efficiently and also contributes to home owner to save energy cost. Because, we know that air conditioning systems consume huge amount of power. If the interior temperature of a home increases, then a air conditioning system also requires more power to cool the inside of a home. Decreasing the gap between the existing interior temperature and the required temperature, an attic ventilation fan contributes to decrease the power consumption of the air conditioning system.

It helps to create a drier and less humid environment in the attic by reducing moisture on the surface. It also reduces the wear and tear and enhances the service life of a roof. Excess heat can cause damage to the roof. Additional ventilation provided by an attic fan can reduce the temperature in the roof area which contributes to enhance roof's lifetime.

LED

What is LED?

Full abbreviation of LED is Light Emitting Diode. LED is a semiconductor device which provides visible light when an electric current is passed through them. LED is a kind of Solid State Lighting (SSL) like Organic Light Emitting Diode (OLED) and Light Emitting Polymer (LEP). After getting an electrical current, the diode emits a bright light around the small bulb. Basically, diodes are used in many technologies like radios, televisions and computers as an electrical component for conduction.

Generally, LED includes common colors like red, green and blue. Characteristically LED is not white light sources. Instead, LED emits closely monochromatic light which makes them highly efficient for colored light applications such as traffic lights and exit signs.

In order to provide white lighting to our homes and offices, LED of various colors are mixed with a phosphor material which converts the color of the light. Phosphor is a yellow material which you can see on some LED products. Now, LED lights are integrated with bulbs and fixtures for general lighting applications.

How does LED work?

Basically, a diode is a semiconductor device which allows current to flow only in one direction. When a diode is connected to an electric current carrying conductor, it excites electrons within the diode and prepares them to release photons which create light. One of the significant features of LED is that it can emit light in a specific direction. Color of the light is a direct result of the energy gap in the semiconductor of the diode. This indicates that LED can easily produce bright spectrum of colors consuming very small amount of power. LED is very sensitive in thermal and electrical conditions. So, it must be carefully integrated into lighting fixtures.

Shapes of LED Lights

LED Lights have five basic shapes. They are as follows:

- **Flood Lights:** These lights are ideal for landscape lighting, outdoor floodlighting and motion sensor.
- **Spot Lights:** These lights are good for track lighting and overhead recessed lighting.
- **Globe Lights:** These lights are used in living room and bathroom vanities.
- **Decorative bulbs:** These bulbs are widely used in wall sconces and decorative fixtures.
- **A-line bulbs:** These bulbs are effective for room area lighting, hallways and reading lamps.

Advantages of LED Lights

LED lights have various types of advantages over other light sources:

- Its efficiency is very high.

- It has high levels of brightness and strength.
- It consumes very small amount of power.
- It can resist vibration and shock. So, it is highly reliable.
- Radiated heat is very low. No Ultra Violet (UV) rays.
- It can be programmed and regulated very easily. It has exceptionally long lifespan.
- It's a cost effective environmentally friendly technology which saves energy and decreases carbon emissions.

Eco Friendly TV

All kind of electronic appliances including television use energy. Typically, a television consumes approximately 75% of power. This power consumption costs a lot of money and it also enhances carbon dioxide emission which is very harmful for the environment.

You should consider purchasing an energy-efficient television which will not only save your money but also contributes to decrease negative impact on the environment.

Eco friendly and energy-efficient television has following advantages:

- Reduces your running costs and energy bills.
- Consumes very small amount of power which helps to reduce carbon dioxide emission.
- Decreases greenhouse gas emissions throughout the functional life of the television.

How to choose a TV

Larger television models consume huge power. Indeed, any larger TV consumes more power than traditional tube models because of their size. Traditional tube models are not even energy efficient. Furthermore, large TV models also consume huge energy during manufacturing, testing, shipping and operating. In order to be more energy-efficient, you need to purchase a model which is no larger than your actual requirements.

Green consumers should purchase Liquid Crystal Display (LCD) television models because of their small power consumption in comparison to plasma sets. A 28-inch traditional cathode -ray tube (CRT) television set consumes almost 100 watts of power. Typically, a 42-inch LCD model can consume almost double power of CRT TV. On the other hand, plasma models consume almost five times higher power of CRT TV. This is why, LCD TVs are considered as the most energy efficient television.

Organic Light-Emitting Diode (OLED) technology is other option to choose a television. In order to create light, this type of screen technology uses organic material. It can produce excellent color, brightness, and contrast. Scientists and technologists are hoping that OLED technology will be a hundred percent energy-efficient in very near future.

Considering only the size and type of screen technology used in manufacturing the television model is not enough while evaluating its energy efficiency. It is also crucial to evaluate its safety level. Consumers should also ensure that the product does not pose fire, radiation or mechanical hazard. This means that the product should not pose any safety risks to the consumer. Remembering this factor, you should check the television model's safety and quality certifications.

Energy Rating Labels or ERL of the television set are also needed to be checked. This is the easy way to choose an energy efficient model. You need to know following matters:

If a television model contains more energy stars then it is considered as more energy efficient.
It is considered to be 20% reduction in energy consumption from the past rating if there is an additional star in the television. For example, a TV model of two star ratings is considered 20% less energy than a TV set of one star rating.

Television units having 7-10 stars ratings are considered as excellent. These ratings of televisions are called "super efficiency rating".

CFL Bulb

What is CFL Bulb?

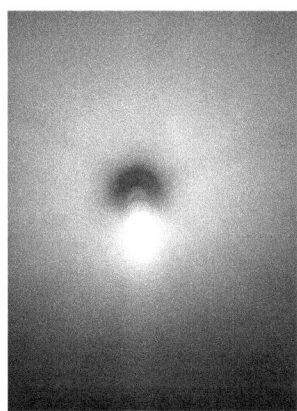

CFL stands for Compact Fluorescent Lamp. In a CFL bulb, an electric current is operated through a tube holding argon and a little amount of mercury vapor. This produces invisible ultraviolet light electrifies a fluorescent coating (called phosphor) inside the tube and emits visible light. During first turning, CFL bulb requires a quite more energy. But, when electricity starts flowing, it use almost seventy five percent less energy than conventional incandescent bulbs. A ballast of CFL bulb helps to start the bulb and then controls the current once the electricity starts flowing. In order to complete the whole process, typically, it requires 30 seconds to 3 minutes. This is why CFL bulbs take longer than other lights in order to ignite completely.

Benefits of CFL Bulb

Efficient: CFL bulbs are more efficient and last longer than incandescent bulbs. A 22 watt CFL bulb has almost same light output like 100 watt incandescent. CFL bulbs use 50 - 80% less energy in comparison to incandescent bulbs.

Cost Effective: Though it is quite expensive initially, it saves money in the long run. Because, CFL bulb utilizes one-third of electricity and lasts up to 10 times as long as incandescent. Using a single 18 watt CFL instead of a 75 watt incandescent will save approximately 570 kWh over its lifetime which saves money. So, CFL bulb is very cost effective and economical.

Reduces Environment Pollution: Substituting a single incandescent bulb with a CFL bulb will contribute to reduce a half-ton of CO_2 from the atmosphere during it's lifetime. If we can use energy efficient lighting like CFL bulb, we could quite 90 average size power plants. Saving electricity means decreasing CO_2 emissions, sulfur oxide and high-level nuclear waste. So, using CFL bulbs contribute to reduce environment pollution.

High-Quality Light: Instead of the "cool white" light of older fluorescent, newer CFL bulb provides a warm, inviting light. They use rare earth phosphors for significant bright color and warmth. New electronically ballasted CFL bulbs don't spark or vibrate.

Flexible: CFL bulbs can be used almost everywhere that incandescent bulbs are used. Energy efficient CFL bulb can be used in lower fixtures, track lighting, table lamps, ceiling fixtures and porch lights. 3-way CFL bulbs having 3-way settings are available now. Dimmable CFL bulbs having dimmer switch are also available in the market.

Rechargeable Battery

A rechargeable battery is one kind of electrical battery which contains one or more electrochemical cells. It's electrochemical reactions are electrically reversible. That's why it is familiar as secondary cell. Rechargeable batteries are of different sizes and shapes. Readily available and most popular batteries that come in often-needed sizes like A, AA, AAA, C, D, and 9-volt are Nickel Metal Hydride, or NiMH. NiMH batteries are a good choice for most battery requires because they have virtually no "memory loss" effect. That means, they do not require to be fully discharged with each use to conserve capacity.

Basically, a battery charger is used to charge rechargeable batteries utilizing grid power. Few others utilize 12-volt DC power outlet of vehicle. While charging, the positive active element is oxidized and produces electrons. On the other hand, the negative element is decreased consuming electrons. Electrons create current flow in the external circuit. In lithium-ion and nickel-cadmium cells, electrolyte works as a safeguard for internal ion flow between the electrodes. In lead–acid cells, it participates in the electrochemical reaction effectively.

Applications of rechargeable batteries

Rechargeable batteries are broadly used in laptops, cell phones, mobile power tools like cordless screwdrivers. In electric cars, electric motorcycles and scooters, electric buses, electric trucks, rechargeable batteries are used as electric vehicle battery. In order to operate under water, submarines use rechargeable batteries. They are used in diesel-electric transmission. For example we can mention, huge trucks, ships and locomotives. In stand-alone power systems and distributed power production, they have significant use.

Benefits of rechargeable batteries

Cost effective: If we can use rechargeable batteries appropriately, it can be used hundreds or even thousands of times. They definitely pay for themselves over time. So, rechargeable batteries are very cost effective.

Conserve resources: As rechargeable batteries can be used over and over, fewer batteries require to be manufactured than with single use diversities. Indeed, rechargeable batteries consume very little non-renewable natural resources than disposable batteries.

Environmental impact: Single use batteries have significant impact on our environment. Heavy metals, destructive materials and other harmful chemicals integrated with inappropriate disposal batteries cause bad impact on our environment. But rechargeable batteries have very little impact on global warming, air pollution, air acidification and water pollution. So, we can say that rechargeable batteries are environmentally friendly.

Performance: Indeed, rechargeable batteries last longer on a single charge than disposable batteries, especially in high-drain devices.

How to handle batteries

- Short circuit creates when positive and negative points of a battery are connected by electric wire. It causes huge damage to the battery or even exploration. So, do not short circuit the battery.
- Do not drop or smash batteries. It can be destructive for cell contents. Do not expose the battery to rain or moisture.
- Always keep batteries away from fire or other sources of extreme heat. Never burn batteries. Exposure of battery to extreme heat may create an explosion.

Solar Lamp

The lamp which is operated by using sun's ray is called solar lamp. Typically, solar lamps are portable. A solar lamp consists of a solar panel and rechargeable battery. During day time, solar panel produces electricity from sun's ray which is stored in the battery. These lights operate at night automatically and shut off when the sun rises in the morning again.

Generally, solar lamp has good outdoor use. For example, we can say that it is used in pathways, garden landscaping and around ponds etc. Basically, solar lamps are attached to a metal or plastic pole, which is fixed directly in the ground. Sometimes, this outdoor solar lamp's top portion resembles a lantern or street light, with a small bulb inside in order to project light.

Typically, an indoor solar lamp emits brightness similar to 40 watts. It can be used as an accent light or desk light. In order to charge fully, these indoor solar lamps are placed in a window which receives direct sunlight for at least six to eight hours each day.

Another type of solar lamp is solar flashlights. These flashlights are slight flat in comparison to a regular one. Basically, it consists of a solar panel in one end which collects sun's ray. In order to provide power to multiple LED bulbs on the flashlight's lens during the evening hours or in dark places, this stored energy is used.

In order to increase the efficiency, little maintenance is required for solar lamps. Few working parts exist in solar lamp. Therefore, the only thing that may need to be replaced is a battery or bulb. Cleaning it with soap and water, we can increase its longevity up to several years.

Unlike grid powered conventional lights, there is no need to be worried about solar lamps being exposed. Typically, the bulb is fixed tightly inside the lamp fixture. It will not get wet even during a heavy shower. Solar lamp's battery is able to resist moisture. These lights are secured to use during an electrical storm. The most significant thing is that they are able to illuminate even if there is a disruption of electrical service.

Retailers comprehensively display these lamps in the spring and summer month when solar lamps are available throughout the year. Generally, they are situated in the landscape segment of lawn and garden or home improvement stores. Solar flashlights should be purchased in outdoor and camping supply stores. These lights are cost effective, fancy and easy to use. It is a significant investment to purchase solar lights for indoor, outdoor and camping use.

Green Holidays

Holidays are important part of our life. When we feel monotonous in our work or personal life, we think to take holidays in order to get rid of stress. Taking holidays are good for our health and mind. During holidays, we enjoy happy moments of our life. We can make our holidays Eco friendly by going green.

Making our holidays 'Green' are significant for our environment and personal life. Green holiday is very spiritual and enjoyable for us indeed.

One of the coolest ways to make a "Green" holiday is arranging thanksgiving dinner around the table with beloved ones, friends and relatives. It's all about simplicity, appreciation and consciousness of many blessings of our life. We can follow following matters in order to make a green holiday with the help of thanksgiving dinner:

- Inviting people for the dinner through electronic mail instead of paper.
- Arrange organic foods, green vegetables, fruits, making green juice with fruits, green tea and green proteins like beans, nut and peas etc.
- Using nontoxic cloth and napkin for the table.
- Non edible table scraps and leaving parts of food can be put on compost pile instead of trash.
- Sharing and enjoying the moment with each other having fun, music, talking and laugh.

Another good example of green holiday is spending time with gardening. It will not only make your holiday enjoyable but also contribute to create a green environment by you which is very significant. During gardening you should use nontoxic gardening materials and organic fertilizer. Leaving parts of gardening can be used to prepare organic fertilizer for future use. Gardening is a green activity and it is very good for your health.

Enjoying your holidays with cycling is a significant example of green holiday. Using cycling as a transport is a good habit. It does not pollute our environment when we use it. For this reason, cycling is considered as a green transport. During cycling, you can enjoy the beauty of green nature which will refresh your mind and also keep your body fit. So, you can easily enjoy your green holiday moment by cycling.

You can also plan to make a green holiday by traveling in some beautiful wild natural places of the world for few days. During the trip, you should choose environmentally friendly accommodation. This type of accommodation should be decorated with non-toxic materials. Required power for the lighting system and water heating system of the accommodation should be generated from renewable energy resources like solar and wind. In this case, solar photovoltaic system, solar thermal system and wind power system can be used. Daylighting can also be used instead of conventional lighting. Heating and ventilation system of the accommodation should be all natural indeed. You can also save energy with your own awareness like switching off lights when you leave the accommodation. During trip, you should use cycle as a transport or you can walk. These are basic examples of green transport. You can enjoy your moment with local people sharing your knowledge and happy moments. You can also build up awareness among local people and encourage them to go green for creating pollution free healthy environment.

Daylighting

Sun is the prime source of light of the earth. Daylighting is the significant way to use that light. Basically, daylighting is defined as the regulated entry of this natural light into a building.

Natural light is justified as 'clean' light in comparison to conventional electric lighting. This natural light is free of electrical flicker. Electrical flicker creates in most florescent lights due to their cycled AC (Alternating Current) power source. In most LED's, it occurs to preserve power. Full spectrum of visible solar colors exists in natural light which is very essential and significant for our eye vision.

A good daylighting system should deliver natural light without glare and should have very little impact on 'envelope' of the building.

Why Daylighting?

The gift of nature is daylight. It is essential for us as like as clean air and water. Our life survives on it. Sometimes we are separated from this basic need.

Natural light has not only artistic benefits of full spectrum lighting but also has significant impact on our health. It helps to maintain our health. It has been researched by researchers that natural light with a balanced spectrum can help maintaining our biological time. Natural light also energizes our body.

Sometimes, due to not having natural light people suffer from sleep depression, sleep disorders and hormone imbalances which cause negative impact on our health, concentration, and performance.

On the other hand, a building with sufficient daylight creates happiness and productivity in our life. These are significantly developed with the help of high quality natural daylight.

High quality daylighting can significantly decrease the use of electric lighting which contributes to save energy and money. It also reduces greenhouse gas emissions.

Benefits of Daylighting

Daylighting has some significant benefits. These are as follows:

- It is observed that daylighting can develop people's lives in many ways. For example, in office settings, daylighting can increase productivity and worker satisfaction.
- Natural light contributes to overcome patient's disease in healthcare. It also develops sleep cycles.
- During study, daylighting increases our learning speed and attention.
- Sun light contains vitamin D which is essential for our body. It is important for the growth of children.
- In factory, warehouse and industrial settings, daylighting delivers higher quality of full-spectrum lighting in comparison to standard HID lights which increases the productivity of workers and also keep the floor environment healthy.
- Daylighting saves energy of the building and improves quality of our life to include daylighting.

Options to include Daylighting

There are lots of options to include daylight into your building space. But, it is necessary to design and plan each option very carefully. For example, skylights and windows are common solutions for simple daylight, but each have their respective disadvantage. Windows provide views and perimeter daylight. They can be good or poor source of daylight depending on building orientation in a particular season and time of the day. Direct sunlight flow through the glass produces bright spots or hotspots which move according to the move of sun. These types of problems also occur with most skylights and even tubular skylights. Sometimes, perimeter or skylight solutions need shades or other systems. It is done by turning off the sun which decreases the benefits to occupants and the operation of the building.

For industrial, commercial and multistory apartment buildings, it is important to plan and design daylighting system very carefully so that it does not decrease building efficiency and occupant productivity. It should also be considered that it is not allowing excessive heat gain or loss.

Geothermal Energy

Basically, geothermal energy is the technique of gaining heat as energy source from earth's surface. In order to produce electricity, most power plants require steam.

Geothermal power plants use steam which is produced from hot water tanks discovered far below earth's surface. A turbine is rotated by steam which helps to run a generator for electricity production. Most power plants are using conventional fuels to boil water for steam. Earth's surface holds huge amount of heat which can produce extremely huge amount of energy than all other natural resources like oil and gas in the world. Predicted heat extraction in geothermal energy is very small in comparison to heat content of earth. So, it is renewable.

Types of Geothermal Power Plant:

Geothermal power plants are of three types. They are - flash steam, binary cycle and small scale power plant.

Flash steam power plants: Here, geothermal tanks of water having temperatures higher than 183°C are used. This hot water flows up with its self-pressure through shafts in the ground. Pressure decreases with its upward flowing and some portion of hot water boils into steam. The steam is then separated from the water and used to rotate the turbine in order to generate power. Any excess water and compressed steam are injected back into the tank which is a significant example of sustainable resource.

Binary cycle power plants: These plants work on water at lower temperatures of about 108°-182°C. In order to boil a working fluid, these plants utilize the heat of hot water which is basically an organic compound having lower boiling point. In the heat exchanger, water fluid is evaporated which helps to rotate a turbine. For reheating, the water is then injected back into the ground. During whole process, working fluid and the water are kept separated. So, there is no possibility of air emissions.

Small-scale power plants: These plants are typically under 5 megawatts. They have good prospect of extensive application in rural areas as well as a distributed energy resource. Distributed energy resource is the diversity of small and flexible power producing technologies which is integrated to develop the process of the electricity delivery system.

Applications of Geothermal Energy:

Long time ago, geothermal energy was used for bathing and space heating. Now, it is known for generating electricity. It can also be used for industrial processes, purification and pumping system.

Benefits of Geothermal Energy:

No fuel is required in geothermal energy and its capital costs are very significant. So, it is very cost effective. It is also environmentally friendly, consistent and sustainable. Historically, geothermal energy is limited to areas close to boundaries of tectonic plate.

Many areas of the world are already using geothermal energy as a reasonable and ecological solution. At present, countries like United States, Japan, New Zealand, Italy, Iceland, Mexico, El Salvador, Philippines, Indonesia and Kenya are generating the most electricity from geothermal sources. This technology helps to minimize the dependency on fossil fuels and also reduces the global warming.

Green Furniture

In the past, during shopping furniture, people typically took decision to purchase based on color, style and price of the furniture. At present, a good number of consumers consider environmental impact of furniture before purchasing it. Furniture elements play a vital role in its carbon footprint. Now, the question is what are the greenest elements to make furniture? The answer is - most furniture is generally made from three elements- wood, metal, and plastic. We can evaluate how these elements associated with health, longevity and environmental impact.

Impact on health

During purchasing furniture for your home, health and safety should be in top priority. Toxins in furniture create poor indoor air quality which causes different kinds of health problems like emphysema, asthma, headaches and fatigue. Volatile Organic Compounds (VOC) from glues, paints, varnishes and adhesives and PVC from finish materials are responsible for the environment pollution of your home. It has been observed that if the product is cheaper, the possibility of more toxic in it is also more.

Look for: Solid wood or metal without VOC finishes are the best options.

Avoid: Fabrics, finishes, paints, fabrics with VOCs and any wood composite materials like plywood and particleboard which are not clearly labelled as 'non-toxic' or 'low-VOC' are likely to be toxic adhesives holder.

Reusing, recycling and life cycle

Positive thinking furniture companies consider the consumer's concern about recycling of products. They always try to design and manufacture their products to be reused or recycled at the end of their life span.

Look for: Metal furniture is often great option in terms of recycling. Steel and aluminum furniture are prepared from recycled content. Sometimes, plastics are also reusable, but it is necessary to check labels in order to be confirmed it before purchasing. Some companies even permit you to return pieces directly to them to be recycled when you are ready to do it.

Typically, it is very complicated to recycle wood, but it can be reused very easily. Also, due to its characteristics of longevity and timeless, you will not ever want to avoid it anyway.

Renewable resources

In order to manufacture furniture, using raw materials from renewable sources is excellent thinking to go green. Renewable products manufactured from bamboo, cork or agro boards are great source of renewable materials.

Look for: Wood is the low impact leader if appropriately harvested.

Avoid: Plastics are manufactured from petroleum. It is very difficult to consider it as a renewable source. Therefore, pieces that are manufactured from recycled plastic should be avoided.

Quality and longevity

The best sustainable choice is building quality indeed. During purchasing furniture, always consider the quality and longevity of materials of furniture in terms of economy and environment safety.

Look for: Craftsmanship and great design are the greenest solution of all. Loving a piece of furniture is as like as keeping it forever and promote it on to next generations.

Avoid: Poorly constructed furniture seems to be a great cheap. It will provide you temporary solution, but there is always a doubt remains in our mind about the longevity and environment impact of these products. So, we should avoid these types of products.

Solar Powered Pump

What is a Solar Powered Pump?

A solar powered pump is a motor pump which is operated by utilizing power produced by photovoltaic module using sun light. The pumping system is capable to draw water from open well or bore well and pond or canal. This pump is very cost effective due to lower operation and maintenance cost. It is a very environmentally friendly technology in comparison to conventional pump which is operated by conventional fuel like diesel, petrol and octane. Solar powered pump is very helpful where no grid power is available.

Types of Solar Powered Pump

Solar powered pump consists of following motor pump sets:

- AC (Alternating Current) surface or submersible pump set.
- DC (Direct Current) surface pump (centrifugal) or floating pump set

DC surface pump is effective where water exists in shallow depth like stream, canals, ponds and open wells. It is operated by DC motor. Maximum suction head is approximately 7 meter and total head is 14 meter. You can get maximum performance of water flow by keeping suction head is kept.

Submersible pumps are suitable where water exists in higher depth. It is a high efficient multistage pump. This pump can lift water from up to 50 meter depth. If AC pump is used then an inverter will be required. Inverter can convert direct current of photovoltaic module into alternating current required for AC pump.

Applications of Solar Powered Pump

Most popular applications of Solar Powered Pump are:

- Agro forestry and plantation.
- Poultry, dairy and sheep farm.
- Aqua culture and fish farming.
- Horticulture farms, vineyards, orchards, gardens and nurseries.
- Supplying drinking water for small inhabitants.

Advantages of Solar Powered Pump

Solar powered pump has following advantages:

- Deliver water during most demand of water.
- No maintenance and labor cost.
- Very convenient to transport, store and remove.

- No fuel cost.
- Long life and reliable.
- No environment pollution.
- It contributes in socioeconomic development.

Wind Power

Wind is the movement or flow of gaseous elements around earth surface. One of the most important characteristics of wind is its velocity. Another characteristic of wind is energy. It is affected by high and low pressure of air of that particular location or area. Earth surface is heated randomly by the sun which depends on the occurrence of sun ray's angle and surrounding of the land. Sun ray's angle can be varied with latitude and time of day. Water of the oceans heat up and cool down slowly in comparison to land. Heat energy which is absorbed in earth's surface is transferred to air directly above these oceans and lands. The density of warmer air is less in comparison to cooler air. These factors are creating constantly changing characteristics of wind across earth's surface. Economically generating wind power from wind is the total amount of economically extractable power available from the wind is significantly higher than power generated from all other sources. Wind energy is a renewable energy which is environment friendly.

What is a Wind Farm?

A Wind Farm or Wind Park is the assembly of wind turbines which helps to generate power from wind in the same area. Typically, a large wind farm has some individual wind turbines which cover prolonged areas. But, the land between the turbines should be utilized for agricultural or other purposes. It may also be constructed in offshore area.

Design of a Wind Farm

There should be a constant flow of non-choppy wind throughout the year having less possibility of stormy wind in the location of a wind farm. Basically, economic wind generators need wind speed of approximately 16 km/h (10 mph) or greater. An important factor of turbine siting is also access to local demand or transmission capacity.

Basically, a wind farm is planned and designed on the basis of a meteorological wind data charts as well as practically measuring actual wind speed of the area. In order to finance the project, it is very important to survey the area and collect the data of wind speed of that particular area. Sometimes, wind in local areas observed for a year or more. Before installing wind generators and detailed wind maps should be prepared.

Each turbine is connected to medium voltage through a combined power system and communications network. In order to increase the voltage of the power system, a step up transformer is used in a substation. Installation of collector system and substation are required to build up a land-based wind farm. A road access for each turbine is also required in the farm area.

Controlling Anger

Anger is a common natural emotion which is related to physiological explanation of upset and mistake of human being. It is very destructive when someone gets out of control. It can be turned into problem when it harms yourself or other persons. Unhappiness and dislike are minor forms of anger. We become angry when we express our reaction of criticism and frustration. Typically, it is a healthy response. Uncontrollable anger creates serious problems in personal relationships as well as work places. It has a significant impact on personal life.

When you lose your temper, you feel and think that it is out of your hands. But you have more control over your anger than you think. Your emotions can be expressed without hurting others. If you are able to implement it you will feel not only better but also you will closely fulfill your basic requirement. Controlling anger is a good practice to overcome angry.

Nature of Anger

Similar to other emotions, it is complemented by physiological and biological changes. Our bodies release the hormones adrenaline and cortisol when we become angry, the same hormones released when we get pressure. Releasing of these hormones raises our blood pressure, pulse, body temperature and breathing rate, often it achieves dangerous levels. This natural chemical reaction is created to provide us prompt increase of energy and power. Sometimes, it is called 'fight or flight' reaction. This means that the body and mind prepare for a fight or for running away from danger.

Sometimes, angry people cannot manage their anger efficiently which causes illness. Naturally, our bodies are not structured to tolerate high levels of adrenaline and cortisol for long time or regular basis. You may become angry at a specific person like coworker or supervisor or event like a traffic jam, a flight cancellation. It can be created due to personal problems like worrying and boring. Shocking or awful memories also causes angry feelings.

Physical effects of Anger

Following changes are happened to your body when you become angry:

- Changing facial expressions
- Feeling tightness in muscles
- Feeling shocked or distressed

How can you control your Anger?

Following practices can help to control angry of human being:

- Breathing profoundly from your diaphragm.
- Gently repeat a relaxing word like "Relax" or "Take it easy".
- Try to memorize funny and happy moments of your life.
- Developing the sense of humor.
- Activity like Yoga can help relaxing your muscles and make your mind peaceful.
- Thinking positively and try to escape saying words like "Never" or "Always".
- Developing problem solving attitude.
- Try to be logical and transform your expectation into your aspiration.
- Increasing better communications with people and society.
- Take leisure after continuous or monotonous activities.
- If you feel that your angry is getting into extreme level, then you can consult with a psychologist or a mental health professional.

Green Finance

Green finance is the concept of integrating the world of finance and business in environmentally friendly manner. Here, individual, business consumers, producers, investors, and financial lenders can participate. Green finance can be promoted in various ways on the basis of participants. It may be managed by financial incentives, aspiration to preserve the planet, or the arrangement of both. Furthermore, it contributes proactively as well as environmentally friendly manner such as promoting the recycling of used products. Green finance avoids any business promotion involves damaging the environment at present or for future generations.

How does it work?

Financial institutions can do it by providing loan to individuals, small businesses and large organizations in an environmentally friendly attitude. For example, loans may be used to support the propagation of renewable energy. A lender should use this loan in order to finance the growth of a solar power plant that generates power from the sun utilizing solar panels. Green financiers also prefer the business of wind power generation. These organizations build up wind farms using large wind turbines onshore and offshore in order to capture wind and generate power.

Power generating companies that use conventional source of fuel like fossil, coal, gas etc. are not eligible to participate in Green Finance. For example, during generating power, coal releases huge amount of emissions in the air which is very harmful and dangerous for environment. As a result, power generating companies that use coal are not eligible to participate for a green finance. Although, clean coal emits very little emissions but may never be categorized as a green investment.

Offering environmental incentives to market participants, green finance may also be encouraged. Green finance has a very proactive format which is allowing small businesses to participate who are not directly engaged in clean energy business. For example, vehicles selling companies concentrated in selling cars which are designed using a hybrid fuel integrated both fossil fuels and renewable energy.

Green energy companies contributing to develop green energy technologies are hoping to generate huge amount of power of world in future.

Green Tea

Green Tea is a tea which is prepared from green leaves of tea tree utilizing lesser oxidation process. As a result, it contains huge antioxidants and helpful poly phenols. Basically, green tea is invented from China. Now days, it becomes popular all over the world. Green tea is the raw material for extracts used in various beverages, health foods, cosmetic items and dietary supplements.

In the past, green tea was used in China and India as a medicine in order to control bleeding and overcome injuries, aid digestion, improve heart, mental health and adjust body temperature. Recently, it is observed by researchers that green tea has positive effects on weight loss disorders of livers and type-2 diabetes.

Nutrition value in Green Tea

Green tea which prepared without any sugar is absolutely a zero calorie beverage. The caffeine contained in a cup of tea can vary according to length of infusing time and the amount of tea infused. Typically, green tea has very little amount of caffeine in comparison to black tea.

One of the healthiest drinks of the world is Green tea which contains huge amount of antioxidants of any tea. Poly phenols are natural chemical which can contribute to provide its anti-inflammatory and anti-carcinogenic effects.

Health benefit of drinking Green Tea

Drinking green tea regularly provides following health benefits:

- Green tea is the excellent source of catechins which is powerful anti-oxidants which contributes reducing the risk for several cancers, including, skin, breast, lung, esophageal, colon, and bladder.
- It is also very helpful in lowering LDL (Bad) cholesterol levels and prevents the abnormal formation of blood clots.
- There is a natural anti-bacterial agent in green tea. During drinking green tea it helps killing the oral bacteria which is the cause of cavities and bad breath.
- Antioxidants in green tea contribute preventing the harshness of rheumatoid arthritis symptoms.
- Green tea is a very powerful against bacteria and viruses of influenza and flu.
- Green tea helps preventing Alzheimer's disease. Alzheimer's disease is a neurological disorder wherein the death of brain cells causes memory loss and mental decline.
- Green tea contributes fighting against allergies.

Green Car

We know that car is one of an important transport of our daily life. But, it should also remember that car is one of prime environment polluter. It pollutes air by emitting poisonous gas and causes damage to green nature of the planet. Green cars are very much environment friendly than conventional cars. Because, they have very less pollutant emissions and use sustainable fuel sources. Basically, it is how the car is fueled which determines whether or not it is a green car.

Why are Green Cars researched and manufactured?

In order to create motion, a conventional car operates with an internal combustion engine which burns huge amount of fuel. This fuel is a fossil fuel and non-renewable. It will be finished one day. Burning of fossil fuel causes the emission of greenhouse gases into the atmosphere. Greenhouse gases are major source of creating global warming.

Fuel is produced from crude oil which is non-renewable resource. The production of crude oil is very dangerous for the balance of Eco-system of the environment. Oil spills is very destructive for natural habitats and kills wildlife.

Considering these factors green car has been developed which causes very less damage to the environment. A green car consumes very little petroleum than conventional cars. There are a number of green cars available today:

Buyers of Green Cars

Green cars are available in the market with wide range. The popularity of green cars is increasing day by day. People are now realizing the necessity of green cars. During purchasing an Eco friendly car that is green car, you should measure the level of emission of CO2 in the air in order to examine how green the car is indeed. Any car having CO2 emissions of less than 100 g/km CO2 is a fairly good car.

Economic benefit of Green Cars

Various types of cars are available in the market at present. So, it is quite difficult to select which car to purchase. In order to help you choosing a car you should consider the fuel efficiency and emissions of a car. A more efficient vehicle is not only good for the environment, but also save your money. Green cars are affordable and cost much the same price as conventional cars.

More fuel efficient cars are small cars compared to large cars. They have very little toxic emissions and your fuel bill will be decreased if the car is driven sensibly.

Every car has their respective fuel efficiency level. The more fuel efficiency means the more environment friendly as well as the more fuel bill saving. Before buying a car, you should check the fuel efficiency by looking at the MPG (miles per gallon). If the MPG is higher, the efficiency level of fuel is also better.

Biophotons

Biophoton is the ultra-weak light which is emitted from biological system. Ultra-weak light is also called photon of light. Biophoton study describes how photon interacts with and within biological systems. Biophoton is a beam of light. It exists in all living substances of this planet. All living cells receive and transmit energy in the form of radiation and light as like as the dependence of life and energy of this planet on sunlight.

Biophotons in food and water support health

In order to sustain and order complex life processes, human cells use this ultra-weak light energy. Nutritious food and vibrant water supply energy to our body. Less entropy or sustained energy means weakening and collapse. Less entropy means greater order, proper metabolic functioning and consistency. Metabolic is the process of breaking down food nutrients into energy. Biophotons in food and water can provide order and energy to the body. Initial benefit of nutritious food is not only for chemical supply of energy, but also for the ability to transport oscillations in the form of biophotons of that food which reduces the entropy of the organism. Entropy is the measure of unavailable energy in a closed thermodynamic system which is also considered as a measure of the disorder of the system. The same methods apply to water and understand the benefit of drinking water.

It has been researched by researchers that the emissions of biophotons from healthy people are significantly higher than ill people. DNA (Deoxyribonucleic Acid), RNA (Ribonucleic Acid) and other forms of macromolecules, including chlorophyll, enzymes and hemoglobin are important sources of biophotons. Wild plant foods are the richest food source of biophotons. These are comestible plants that have grown without any involvement of humans. Wild plants consist of more nutrients such as vitamins, minerals, antioxidants, photo-chemicals, and enzymes than cultivated organic crops. Researchers also experimented that cooked or irradiated food emitted virtually no biophotons. After wild plants, the next best sources of biophotons are freshly picked fruits and vegetables which is grown organically.

It is also observed by researchers that the biophotonic radiation from fresh fruits and vegetables contain recommended amounts of all vitamins as like as multivitamin.

In order to get the highest level of biophotons from wild plant foods and other organically grown foods, eat a diet rich in raw fruits and vegetables which is picked from your own organic garden freshly.

Green Jobs

What is Green Jobs?

Jobs which contribute to improve the environment and reduce the destructive impact on earth are called Green Jobs. That's why environmental conservation, recycling and pollution reduction, green manufacturing, renewable energy (solar, wind, geothermal, biomass etc.) and constructing green building are sectors of green jobs. Green jobs contribute to create good impact on environment.

Why Green Jobs?

We know that conventional source of fuel like gas and oil will be finished one day. So, we should think what will happen then and what the alternative solution is. In this scenario, renewable energy sources like sun, wind, biomass and geothermal are good solution. Moreover, this technology is environmentally friendly. We are well aware of climate change and global warming. These are happening due to carbon emission in the air which is created for following reason.

- Industries are releasing carbon dioxide gas in the air during manufacturing of their products.
- Releasing carbon monoxide and other harmful gas in the air by transports like bus, car etc.
- Cutting huge amount of trees to manufacture furniture and other commercial purpose which creates deforestation.

In order to solve these problems and save conventional fuels like oil and gas for future use, we should move forward to green energy. Demand and field of green jobs are increasing day by day. Organizations involving in green energy sectors are creating green jobs which are very important in the context of our environment and economy. We can contribute in this field by creating green jobs.

What skills are required for Green Jobs?

Skills for green jobs are often the similar skills required for traditional jobs. If you can able to transfer these traditional skills to a green project or process, you can launch into green. Following skills are essential to apply for green jobs:

Mechanical skills: In green jobs, mechanical skills and aptitude are needed for skilled manual work. For example, installing, troubleshooting and operating of mechanical and laboratory testing equipment of green energy sector like solar, wind, biomass and geothermal etc.

Math skills: Basic math skill like recording expenditures, preparing drawings and specifications, operating programmable control tools, and collecting quantitative data are needed. Advanced math skills are required in statistical analysis and engineering calculations.

Technical skills: In order to evaluate, design, install, operate, monitor, or rectify malfunctions of anything from an air duct to a technology-intense environmental compliance program, technical skills are required wherein theoretical knowledge such as scientific or legal is very crucial.

Design skills: For developing new green products or find new ways of making existing products or processes more environmentally sustainable design skills are essential. Workers with a high level of creativity, problem solving, and critical thinking skills combined with technology design skills are needed for the green economy. These skills are developed gradually through exclusive training and experience in technical fields like engineering.

Research and analysis skills: Scientific investigation, compliance auditing, data collection, quality control analysis, and the accurate reporting of findings are required research and analysis skills.

Project management skills: Initiative of green activities is created in the form of projects like site remediation projects, sustainable building projects and conservation planning projects. Project management skills are very crucial for any industry including green energy sector. Some of these skills are - defining the project's financial, market and business goals, communicating and coordinating across all project stakeholders, tracking and adjusting plans when necessary, quality review, document preparation, management of material resources, financial resource management, time management and personnel management.

Sales skills: Technical knowledge of green products and the development of long-term relationships with the customer are required sales skills. In order to be successful in selling products, you need profound interest in listening attentively and working closely with the customer to identify new product/market opportunities and customized solutions.

Green IT

What is Green IT?

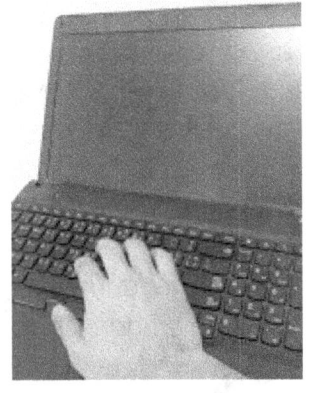

Green IT (Green Information Technology) is the technological development of environmentally friendly sustainable computing or IT which includes designing, manufacturing, using and arranging of computer, servers, and related subsystems such as monitors, printers, storage devices, and networking and communications systems efficiently and effectively with lesser or no impact on the environment. A typical computer consists of thousands of chemicals and if we put all the computers in the world together, they utilize huge amount of electricity. Goals of green computing are to reduce the use of hazardous materials, maximize energy efficiency during lifetime of the product and promote the biodegradability of during the product's lifetime, and promote the biodegradability of obsolete products and factory waste.

A computer contains over one kilogram of lead and different type of chemicals which includes antimony, arsenic, boron, phosphor, nitric acid, hydrofluoric acid, and hydrogen fluoride etc. Electronics contain copper. This toxic and inefficient cocktail is where the term e-waste (electronic waste) has been created. Another reason is we have limited resources on earth. When we use minerals like gold and silver and we don't recover them, those minerals are deducted from our overall supply. Eventually, very few totals will remain. These resources will no longer available to us if we don't recycle what we use.

Energy can also be saved through cloud computing, namely the principle of outsourcing the programs and functions of one's own computer to service providers over the internet. It is also called sharing storage capacity with others. In this case, smaller devices, some purely mobile, are all that is required to handle large volumes of data.

Another idea is the joint use of purely virtual space. 'Digital sharing' is the way of putting it. New ideas and programs for video conferences are presented here in order to prevent people from taking unnecessary flights and to be climate-friendly by shifting business meetings to their own desk. Energy can be saved by using a laptop instead of a large desktop computer at workplaces or in home offices. Because, laptop components are energy optimized, primarily to ensure that the battery lasts longer.

Bringing modern LCD monitors into the picture is also a better idea of Green IT. Because, traditional CRT monitors have particularly high energy consumption. A classic green IT tip is to abandon the power-guzzling stand-by mode. We now have a range of different electrical sockets with on-off switches that should be deployed when using electronic devices for longer periods.

IT Managers can monitor energy requirements and can cap power usage and place servers in power saving mode with server power management. We can also apply green storage technique. For example, instead of running database that requires 5000 IOPS on 250 disk drives, put the heaviest hit portions of that database on a pair of mirrored and protected solid states disks. The rest of the data can use a smaller number of traditional fast disks while backups can go to high capacity SATA drives.

Enterprise desktop power management software automatically places monitors and computers into low power modes-sleep or hibernation. Using blanking panel in the data center infrastructure is another example of green IT. Going paperless plays an important role in green IT. Using more papers to print documents requires additional power consumption and ink of printers which creates additional cost and also it has significant impact on earth. Because, in order to produce papers, paper industries require massive power consumption and emission of greenhouse gas. So, reducing the use of paper and increasing the use of soft copy for documentation can contribute developing Green IT significantly.

Green Carpet

Indoor pollutant levels are two to five times higher than that of outside. In order to find the source of many of these pollutants, just observe down. When you install a new carpet and flooring in your house or office, you also install hundreds of volatile organic compounds (VOCs), including known and suspected chemicals like formaldehyde and benzene in the air automatically. In order to remove these components, long time even years are required. Moreover, carpets are often treated with toxic chemicals for moth proofing or to hold off soil and moisture. Carpeting is also dangerous for trapping toxic lawn chemicals, VOCs, and allergens tracked in from outside.

In order to minimize indoor pollution and reduce health problems created by toxic carpets, sustainable flooring is a great option. Rapidly growing line of carpets and flooring made from recycled and Eco friendly materials, you can choose now. These carpets are called green carpet. They are durable, stylish, and cost effective than conventional floors and carpets. These eco-friendly green carpets provide a responsible and healthy way to enrich your home.

Rolling Out the Green Carpet

From the beginning of purchasing to the end of the disposal of carpet, environmental and health costs are incorporated with carpeting. Here are some techniques to minimize those costs:

Fast and cheap fix: If you are experiencing health problems that could be caused by your carpet, there is a cheap solution. Try a nontoxic, green carpet.

Carpet: Purchase carpets made from natural fibers with little or no chemical treatment and natural-fiber backing attached with less-toxic adhesives.

Padding: Many carpets and carpet padding have plastics which are manufactured from an nonrenewable and energy consumptive resource petroleum. A carpet with lightweight backing that requires no additional padding, or use padding made from recycled materials should be chosen to purchase.

Installation: Your journey toward eco-friendly floors start when you search the right carpet. Next, you need to install your carpet installed. It is a process which often involves chemical-based glues associated with respiratory problems and other health issues. In order to gluing that eliminates many potentially hazardous pollutants, it is safe and easy to track carpets down. However, if you do decide to glue, you can take steps to minimize your ecological footprint. Try to use water-based, low-VOC glues to install your carpets.

Green Cosmetics

Cosmetics which are manufactured using all natural, non-toxic ingredients are called green cosmetics. It is also known as bio cosmetics or Eco-friendly makeup. Naturally occurring mineral ingredients are used by many green cosmetics for coloring and sun protection purposes. Basically, these products are manufactured in an environmentally sustainable method. During marketing of these green cosmetics, environmental sustainability concern is considered.

Face makeup, such as foundation, cosmetic powder, eye shadow, rouge, lipsticks, and lip glosses etc. are examples of Eco-friendly makeup and beauty products. Besides, hair products, such as shampoos, conditioners and styling gels or mousses, lotions, lip balms, facial masques, soaps, cleansers, and even toothpaste are produced by manufacturers of green cosmetics. Most of cosmetic or hygiene product used on the skin or hair is available in a green form.

Plant extracts and oils, naturally occurring minerals, non-toxic dyes and natural ingredients are used during manufacturing of these products. Butter, rose hips, mica, lemon, coffee, and carrot are some additional ingredients commonly used in these products. It's usually possible to prepare one's own green cosmetic products, since the ingredients are typically available and affordable. Everyone should carefully read the ingredients list on any manufactured green cosmetic product in order to confirm that there is no synthetic chemical in the product.

Some other examples of green cosmetics are lipsticks, mascaras, and facial powders which enrich the appearance. These products can also be useful for skin care and health. Lotions, soaps, facial masques, and other products that can exfoliate, moisturize, and soften the skin are available. In order to minimize the signs of aging, some manufacturers of green cosmetics have produced products designed. There are mineral ingredients in some products which can help prevent sun burn and sun damage, while others contain vitamins and nutrients said to support the health and vitality of the skin.

Walking for Health

Walking is an effective exercise for everyone to diminish their health risks and develop their body fitness. Regular walking has following benefits:

- Walking reduces the risk of high blood pressures and cholesterol.
- It helps controlling diabetes diseases.
- Everybody can control their weight through regular walking.
- Bones become strong and balanced.
- Muscle strength and stamina will be enhanced.
- Excessive fat of the body will be decreased.
- It helps improving vitamin D levels.
- It contributes reducing you depression, better sleep, and more.

Walking daily for 30 minutes

In order to get maximum health benefits, everybody should try to walk for at least 30 minutes as sharply as you can on most days of the week.

Stretching before Walking

In order to prepare the joints and muscles for the improved level of motion required, slight stretching is very essential prior to start walking. To warm up the muscles before stretching, take an easy five minute walk. Consult with a healthcare practitioner the proper way of doing stretches, and be confirmed to include the neck, arms, hips, upper and lower leg muscles (including the hamstring muscles in the back of the thigh), and ankles.

Routine wise walking

Try to grow up the habit of walking routine wise. That is walking every day at the same time. Same amount of energy should be used during walking; no matter what time of day you walk, so do what is most suitable for you. Request someone to walk with you which will help make a regular activity for you. Keeping an activity diary or log also makes it someone it easier. Building up the habit of regular walking will give us a positive result to our health.

Walking Techniques

Following techniques will help developing the benefits of walking:

- Walk sharply. But, as a common rule, you should maintain enough breath in order to be able to carry on a conversation.
- Begin with a 5 minute walk and continue walking for at least 30 minutes (roughly 2 miles) at least 3 to 4 times a week.
- During walking, you should maintain good form in order to get maximum aerobic (requiring air) benefit with each step and help protect the back and avoid injury.

Following elements of form should be followed:

Head and shoulders: Head should remain up and centered between the shoulders focusing eyes straight forward horizontally. Shoulders should be relaxed but straight - avoid lounging forward.

Front muscles: For supporting the chest of the body and the backbone, the front muscles should use effectively. In order to do this, you should slightly keep the stomach pulled and stand fully straight. During walking, tilting forward should be avoided.

Hips: Begin forward motion with the hips. Each step should feel natural - not too long or too short. Don't try to take too long of step.

Arms and hands: Fixing elbows at a 90 degree angle, your arms need to stay close to the body. During walking, your arms should keep in motion, swinging front to back in pace with the step of the opposite leg. You should keep your hands relaxed, lightly cupped with the palms inside and thumbs on top. Tightening the hands or making tight bunch should be avoided.

Feet: Land slightly on the heel and mid foot with every step, progressing smoothly to push off with the toes. Carefully, use the balls of the feet and toes to push forward with each step.

Walking Shoes

Walking shoes give a basic protection and mechanical support for the base structure of body. It keeps the entire body balanced. Imbalance in the feet can cause changes in the body. So, we should use correct shoes during walking.

Using Correct Walking Shoes

Correct walking shoes are very crucial for excellent balance during exercise walking. Because, pain or injury can be created due to use poorly fitted walking shoes. Technical running shoe store can provide a shoe that fits based on the individual's specific bio mechanics.

Some shoes have the design pattern of outward moving and some shoes have inward moving design patter. So, it is very essential to choose walking shoes that match each individual's specific bio mechanical pattern.

Purchasing Walking Shoes

Before purchasing a new pair of walking shoes following crucial factors should be considered:

- **Firmness** - A stable and confident feel throughout range of motion are the basic requirements of shoes.
- **Elasticity** - Shoes should provide smooth motion in order to get a good degree of give at the base of the toes.
- **Comfort** - Walking shoes should contain shape and padding adapted closely to the feet, providing a warm fit at the heel and mid foot having sufficient room in the forefoot.
- **Weight-** Walking shoes should be light weight and breathable.

Green Building

A green building is the integrated structure of design, construction and operational practices that significantly removes its negative impact on the environment and its tenants. It is an opportunity to use resources efficiently while creating healthier environments for people to live and work in. Green building can also reduce construction and performance costs significantly.

Green buildings leave a lighter footprint on the environment through saving resources. It also balances energy-efficient, cost-effective, low-maintenance products for construction needs. That means, green-building design is engaged to find the delicate balance between home building and a sustainable environment.

What makes a building "Green"?

A green building is responsible environmentally and it is resource -efficient throughout its life-cycle. These characteristics develop and balance the classical building design in the context of economy, utility, durability, and comfort.

In order to reduce the overall impact of the built environment on human health and the natural environment, green buildings are designed as follows:

- Using energy, water, and other resources effectively.
- Keep the health of tenants safe and increasing employee efficiency.
- Decreasing waste, pollution and environment degradation.

For example, green buildings may integrate sustainable materials in their construction (e.g., reused, recycled-content, or made from renewable resources); creating healthy indoor environments with reduced toxin (e.g., reduced product emissions); and/or feature landscaping that reduces water usage (e.g., by using native plants that survive without extra watering).

What are the benefits of green building?

Green buildings have a huge impact on the environment, human health, and the economy. Successful implementation of green building policies can maximize both the economic and environmental performance of buildings. However, the most significant benefits can be obtained if the design and construction team takes an combined method from the earliest stages of a building project. Potential benefits of green building can include:

Environmental benefits

- Increase and protect biodiversity and ecosystems
- Develop air and water quality
- Diminish waste streams
- Save and restore natural resources

Economic benefits

- Minimize operating costs
- Generate, increase, and outline markets for green product and services

- Develop tenant productivity
- Enhance life-cycle economic performance

Social benefits

- Enrich tenant comfort and health
- Amplify visual qualities
- Reduce stress on local infrastructure
- Enhance overall quality of life

How do buildings affect climate change?

In order to heat and power our buildings huge amounts of energy is used. Here, the source of energy is burning fossil fuels - oil, natural gas and coal - which generate significant amounts of carbon dioxide (CO2), the most extensive greenhouse gas.

Decreasing the energy use and greenhouse gas emissions produced by buildings is therefore essential to the effort to reduce the speed of global climate change. Buildings may be linked with the release of greenhouse gases in other ways, for example, construction and destruction debris that destroys in landfills may generate methane, and the removal and manufacturing of building materials may also generate greenhouse gas emissions.

Biofuels

Biofuels are fuel having energy from geologically recent carbon fixation which are produced from living organisms. Carbon fixation exists in plants and micro algae. These fuels are produced from a biomass conversion.

There are various kinds of biofuels in many countries of the world. For decades, Brazil has converted sugarcane into ethanol. There, some cars can run on pure ethanol instead of fossil fuels. In Europe, a diesel like fuel biodiesel is produced from palm oil.

Benefits of Biofuels

Biofuels are promising potential great solution. Cars are a major source of atmospheric carbon dioxide, the main greenhouse gas causing global warming. As plants can absorb carbon dioxide with their growth, crops grown for biofuels should suck up about as much carbon dioxide as comes out of the tailpipes of cars that burn these fuels. And unlike underground oil reserves, biofuels are renewable resources. We can always grow more crops to convert it into fuel.

It may be a better idea to produce biofuels from grasses and saplings wherein more cellulose exists. Cellulose is a tough material which constructs plants' cell walls and most of the weight of a plant is cellulose. It could be more efficient than existing biofuels and emit less carbon dioxide if cellulose can be converted into biofuel. In three different ways, biomass can be converted into convenient energy. They are - thermal conversion, chemical conversion and biochemical conversion. It can produce fuel in the form of solid, liquid or gas. This new biomass can be used for biofuels. Due to possibility of increasing oil price in future and the demand for energy security, the popularity of biofuels is increasing gradually.

We can use biodiesel as a fuel for vehicles in its pure form. But, typically, it is used as a diesel in order to reduce carbon monoxide, hydrocarbons and other harmful elements from vehicles which are operated by conventional fuel diesel. Biodiesel is produced from oils or fats using transesterification (Process of exchanging between organic groups) and is the most common biofuel in Europe.

Green Recipe

There are many advantages to start your day with detoxifying green drinks. Some include clearer skin, deeper sleep, less stress, increased energy, better digestion, and improved circulation. This is because you are putting cleansing greens into your body that are easy on the digestive system. The easier your food is to digest, the more energy your body can use towards getting harmful toxins out of your body because it isn't using that energy for digestion. That means your body will have time to fight the cause of your spoiled skin and the toxins causing it will be flushed out. These detoxifying green drinks are so delicious that you will certainly enjoy doing your body a favor.

1. Stomach Cleanser Green Smoothie

This is the strongest smoothie among all the detoxifying green drinks. Mint is a powerful blood detoxified element which reduces bloating. Lemon detoxes the blood as well, but it also flushes toxins from the liver and kidneys and aids in digestion. The taste of refreshing mint pairs nicely with the lemon in this smoothie, creating a light and delicious green-colored drink.

Ingredients:

- 3 cups water
- Juice of 1 lemon
- 1 inch piece of ginger
- 4 inch piece of cucumber
- ¼ cup mint
- 1-2 handfuls ice

Directions: Blend all of the ingredients in a blender and blend until smooth.

2. Spinach smoothie with Avocado and Apple

Serves 2.   Hands-On Time: 05m.   Total Time: 05m

Ingredients:

- 1 1/2 cups apple juice
- 2 cups chopped and stemmed spinach or kale
- 1 apple—cored, unpeeled and chopped
- 1/2 avocado, chopped

Directions: Blend the apple juice, spinach, apple, and avocado in a blender and continue blending until smooth. Continue it for approximately 1 minute. In order to get required consistency adds some water.

Nutritional Information

Per Serving

- Calories 244
- Fat 7g
- Saturated Fat 1g
- Cholesterol 0mg
- Sodium 33mg
- Protein 3g
- Carbohydrate 42g
- Sugar 28g
- Fiber 6g
- Iron 2mg

- Calcium 100mg

Power of Meditation

Meditation is a practice of transforming the mind. Meditation practices are techniques that encourage and develop concentration, clarity, emotional positivism and a calm seeing of the true nature of things. By involving with a particular meditation practice you learn the patterns and habits of your mind, and the practice offers a means to cultivate new, more positive ways of being. With regular work and patience these nourishing, focused states of mind can deepen into profoundly peaceful and energized states of mind. Such experiences can have a transformative effect and can lead to a new understanding of life.

In order to obtain a perfect state of health, one has to remain mentally calm, steady and stable. Health is not just confined to the body and the mind but it is also connected with the consciousness. The clearer the consciousness is, the more well-being is gained.

Meditation increases life energy

The vital life energy is the very basis of health and well-being, for both body and mind. You can gain this through meditation. When your body is alive with more life energy, you feel alert, energetic, and full of good humor. A lack of life energy results in lethargy, dullness and poor enthusiasm.

Dealing with illness - through meditation

The root of an illness is in the mind/consciousness. So, by attending to the mind, clearing it of any disturbances, the recovery speeds up. Illnesses can develop from:

- **Violating natural law:** such as over-eating
- **Imposed by nature:** such as common cold, an epidemic

Nature itself gives us a cure for these illnesses. Health and illness are a part of physical nature. By practicing meditation, the stresses, worries, anxieties drop off and gives rise to a positive state of mind, which has a positive impact on the physical body, brain and nervous system, then illnesses change.

Health and illness are a part of physical nature. You should not worry too much about it.

When you worry about illness, you are giving more power to the illness. You are a combination of health and illness. When you keep that in mind and have a positive state of mind, then illnesses change.

Heal the mind – through meditation

Meditation helps removing stresses from entering the body system and also releases accumulated stress. Health, happiness and a positive state of being wells up. Meditation brings a steady state to the brain; it's like servicing the whole body-mind complex.

Clear emotional pollution - with meditation

People around you, affect your state of mind. They either give you peace and joy, or create disturbance (such as jealously, anger, frustration or sadness). You're affected because the mind is not in its Self; it is not centered. Meditation is the key to control 'emotional pollution'.

Blossom with meditation

Meditation brings spiritual transformation. As you learn more about life, the mystery of the whole Creation unfolds. Then the questions that arise in the mind are - What is the meaning of Life? What is its purpose? What is this world, what is love, what is knowledge?

After raising these questions in you, you will know that you are very fortunate. These questions need to be understood; you cannot find the answers in books. You have to live through them and witness the transformation. That is perfect health; you are transformed from within. And the bud becomes a totally blossomed flower.

Heal the world through meditation

Meditation filters the environment. Meditation has transformed aggression and violence in people - to compassion, love and care. For example, notice how you feel when you enter a room where someone is really angry? It leaves you feeling the same!

Similarly, when there is a happy activity, you feel good. You may wonder why. Feelings are not isolated in one's body - they are all around. It is in the whole environment, because the mind is subtler than the five elements (earth, water, fire, air, ether). If there is a fire somewhere, the heat is not just in the fire, it is also radiating throughout the place.

If you are unhappy or depressed, you are not the only one who is feeling it; you are spreading it to the whole environment.

In the present global situation of conflicts and disease, it's important to meditate a little every day. Through meditation, you can nullify the negative vibrations in the environment, thereby creating a more harmonious environment.

The Healing Breath and meditation

This unique breathing practice is a potent energizer:

- Every cell becomes fully oxygenated and flooded with new life.
- Negative emotions from the body will be flushed out.
- Tensions, frustrations and anger will be released.
- Anxiety, depression and lethargy will disappear.
- Both the mind and body will be relieved.

After finishing the practice, one is left calm and centered with a clearer mind. A sense of joy in the moment prevails and your heart will encourage you to smile.

King of Herbs

Which herb is the king of herbs?

Basil herb is the king of herbs, which is one of the oldest and popular herbal plants having health benefiting element photo-nutrients. This highly prized plant is known as "holy herb" in many traditions all around the world.

Basil is of the family of Lamiaceae, of the genus: Ocimum. Its scientific name is "Ocimum basilicum". Asian or "holy" basil (Ocimum sanctum) has large, hairy plant with pink flowers and pink leaves. Mediterranean sweet basil (Ocimum basilicum) has light green leaves.

Basil is originated from Iran, India and other tropical regions of Asia. This bushy annual herbal plant has useful leaves and seeds having medicinal value. Basil grows well in warm and tropical climates. On an average, matured plant reaches the height of 100 cm. Its leaves are light green, silky about 2.5 inches long and 1 inch broad with opposite arrangement. The flowers are quite big, white and arranged in a terminal point.

Health benefits of Basil herb

- Powerful chemical compounds of basil leave have disease preventing and health improving characteristics.
- Poly phenolic of basil herbs has anti-oxidant protection power against radiation-induced lipid per-oxidation in mouse liver.
- Health beneficial oils exist in basil leaves having characteristics of anti-inflammatory and anti-bacterial.
- There are very low calories in the part of basil herb having no cholesterol. It contains nutrients, minerals, and vitamins which is vital for our health.
- Huge amount of beta-carotene, vitamin A and some other vital elements exist in basil herb. These compounds are very powerful against oxygen-derived free radicals and reactive oxygen species (ROS) which protect us from aging and various disease processes.
- A yellow element of basil herb protects our retina from UV (Ultra Violet) rays. It is observed that common herbs, fruits, and vegetables having huge amount of anti-oxidant element contributes to protect us from age-related diseases, especially in the elderly.
- There are 5275 mg or 175% of vitamin A exist in 100 gm of fresh herb leaves. Vitamin A has antioxidant properties which is essential for our vision. It is vital for maintaining healthy mucus membranes and skin. It can protect our body from lung and oral cavity cancers.
- Basil herbs contain vitamin K which is very vital for many coagulant factors in the blood. Coagulation is a change to a sticky, jelly-like state. It contributes to strong our bone function with mineralization process.
- A good amount of minerals like potassium, manganese, copper, and magnesium exist in the basil herb. Potassium is an important component of cell and body fluids, which helps control heart rate and blood pressure. Manganese is used by the body as a co-factor for the antioxidant enzyme.

- A good source of iron is basil herbs. Per 100 gm of fresh basil leaves have 3.17 mg of iron. Iron exists in hemoglobin of our red blood cells. It can regulate the oxygen-carrying capacity of our blood.

Green Energy

Green Energy is the energy which is produced from natural resources like sunlight, wind, tidal wave and geothermal heat. It is also called renewable energy or alternative energy and sustainable energy. Green Energy is extracted, generated and consumed without any significant negative impact to the environment. It does not add any CO_2 in the atmosphere.

Why Green Energy?

Energy demand in this planet is increasing day by day. Green power is produced from renewable energy resources which provide the highest environmental benefit and impact to us. Green energy is a very environment friendly technology and it helps reducing carbon emission which is very significant for our planet. It replaces conventional fuels like oil, gas and so on. Conventional fuel can be finished in near future. So, we have to think about the solution of it. In this scenario, alternative energy or green energy can keep some contribution in terms of future energy context. It contributes in generating electricity which is very important in terms of economic and environmental context. It delivers power from solar, wind, geothermal, bio gas, biomass, and low-impact small hydroelectric sources.

How can we get Green Energy?

We can get green energy from our environmental source like sun, wind, bio gas, biomass and geothermal. One of most popular way to get this green energy is installing PV system.

What is PV system?

PV system stands for photovoltaic system. This is a technology wherein electricity is generated from sun without polluting environment. For residential, commercial and industrial energy supply, a PV system basically consists of following components:

- PV module.
- Battery.
- Charge Controller.
- Inverter (DC to AC converter).
- Mounting structure for PV module.
- Electrical wiring and interconnections.

PV system may be constructed in different configurations. They are as below:

- Off grid without battery (array-direct).
- Off grid with battery storage for only DC appliances.
- Off grid with battery storage for both AC and DC appliances.
- Grid tie without battery.
- Grid tie with battery storage.

Carbon Trading

Carbon trading is the practice of reducing overall emissions of carbon dioxide, along with other greenhouse gases, by providing a regulatory and economic incentive. Indeed the term "carbon trading" is a quite misleading, as the number of greenhouse emissions can be regulated under what are known as cap and trade systems. As a result, some people prefer the term "emission trading," to emphasize the fact that far more than just carbon is being traded.

Benefit of Carbon Trading

It provides a very good incentive for companies to improve their efficiency and reduce their greenhouse gas emissions, by implementing such reductions into a physical cash benefit. Moreover, it is a disincentive for being inefficient, as companies are effectively penalized for failing to meet emissions goals. In this way, regulation is accomplished largely through economic terms, rather than through draconian government measures, encouraging people to engage in carbon trading because it's potentially profitable. Basically, it is a practice which is designed to reduce overall emissions of carbon dioxide, along with other greenhouse gases, by providing a regulatory and economic incentive.

Carbon credits are effective solution to reduce the amount of Green House Gas (GHGs) emissions in the atmosphere. Generating and selling carbon credits funds carbon projects which would not have gone ahead i.e. additional to business as usual. Carbon credits also contribute lower the costs of renewable and low carbon technologies as well as assisting in the technology transfer to developing countries.

Implementation of Carbon Credits

Carbon credits can be implemented from various types of projects including:

- **Renewable energy:** Shifting from fossil fuels to a 'clean' energy. For example, wind and solar energy.
- **Forestation and Afforestation:** New tree plantation contributes absorbing CO_2

by trees. For example, forest regeneration.

- **Energy efficiency:** Decreasing emissions through increasing energy efficiency. For example, installation of energy-efficient machinery.
- **Methane capture:** Avoiding methane emissions through capture and burning to create energy. For example, landfill methane capture.

The eligibility of the project for carbon credits depends on whether a project follows one of the Kyoto Protocol's project-based mechanisms or an independent voluntary standard.

Government sets a national goal for total greenhouse gas emissions over a set period of time, such as a quarter or a year, under a cap and trade system and then allocates "credits" to companies which allow them to emit a certain amount of greenhouse gases. If a company is not able to use all of its credits, it can sell or trade those credits with a company which is panic to exceed its allowance.

Jute Industry

Jute is a natural fiber. It is obtained from the bark of the jute plant as an extract that grows like any other organic crop. In early days, this fibrous plant was also used by the occupants as a fragility that went along with their staple diet.

Beginning of jute as a commercial product dates back to the end of the 18th century. Initially, the jute fiber was made into ropes that were extensively used in the wind and hand driven sea vessels and ships. Later, jute was used as spun and woven for manufacture of carpets.

Newer technologies developed within 1838. For manufacturing jute cloth, jute fiber was spun into better yarn and woven. In order to enter the daily lives of the people, jute products were initially developed with sacking bags and jute hand bag. Jute also leads the applications in carpet making and packaging since jute started to be woven into fabric form.

Some other properties of jute fiber started to develop since the middle of the last century. It appeared then, that jute fiber and its subsequent processing might find application in new areas of use and also newer products for consumers.

The best alternatives of non-bio degradable plastic bags are jute products, which becomes popular in many countries. Plastics have raised environmental hazards, blocking of drains and natural water streams and many more dangers. Jute bags and paper bags are thus gaining popularity for a good cause. The alternative may not come as cheap as its plastic counterpart, but the price paid will still be cheap for the cost of saving to the environment.

There is a good demand of jute products in countries like USA, U.K., Germany, Australia and Middle-east. Demand of jute products in these countries are jute shopping bags, wall hangings and floor coverings. New markets can be developed through consumer awareness and product promotion it will not be surprising to find this natural fiber product become a much of the material for regular use by consumers all over the world.

Jute fiber finds its use in the fields of agricultural, industrial, commercial and domestic. Sacking and Hessians (Burlap) constitute the bulk of the manufactured products. Sacking is commonly used as packaging material for various agricultural commodities viz., rice, wheat, vegetables, corn, coffee beans etc. Sacking and Hessian Cloth are also used as packing materials in the cement and fertilizer manufacturing industries. Fine Hessian is used as carpet backing and often made into big bags for packaging other fibers viz. cotton and wool.

India, Bangladesh, China, Thailand, Myanmar & Nepal are the major producing countries of Jute and Allied Fibers (JAF). Around 95% of the global production of JAF is produced together by them. India and Bangladesh produce mostly jute; China produces mostly kenaf while Thailand produces kenaf and roselle.

Basically, the practice of fermenting the jute plants in jute growing regions is to engage the jute bundles in clear slow flowing water, in canals, streams, tanks and ponds. Plant materials of jute to water should have at least 1:20 ratio in still water.

Water should be free from salt and clear. The volume of water should be enough to allow jute bundles to float. During submerging, bundles should not touch the bottom. The same fermentation tank should not be used when water becomes duller. In order to extract fibers from jute and allied vegetable fiber plants, fermentation process are applied for a long time. As the fibers are contained in the bark or the outer skins of stems, either stems or the outer skins called ribbons are fermented for extracting the fibers. When stems are fermented, then it is called stem fermentation. When ribbons are fermented, then it is called ribbon fermentation. Fermentation is a critical step to produce good quality fiber.

The quality of jute fiber is evaluated by its appropriateness for the production of various types of yarn and its behavior in the manufacturing process. The fiber which spins into the finest yarn is considered to be of very good quality.

Jute fiber is marketed in bundles of fiber balls. A fiber ball is composed of about 10-15 fiber reeds obtained from 10-15 plants. Each fiber reed is composed of thousands of fiber strands made of ultimate fibers with strengthening materials. Commercially fiber quality is assessed by taking a ball out of a lot, scattering the individual canes on the ground and then assessing the different characteristics by 'look & touch' method.

Jute is an annually renewable energy source with a high biomass production per unit land area. It is biodegradable and its products can be easily prepared without causing environmental hazards. The roots of jute plants play a vital role in increasing the fertility of the soil. By revolving with other crops like rice and potatoes, jute acts as a barrier to pest and diseases for others crops. It also provides a significant amount of nutrients to other crops as a form of organic matter and micro nutrients. Jute has ecological flexibility. It can be grown on a variety of soil types. They have a good acceptance to salinity, water stress and water logging. Jute has the resistance ability to climatic extremes, pests and diseases.

Carbon dioxide (CO_2) absorption rate is huge in jute plants. It can clean the air by consuming large quantities of CO_2, which is the main cause of the greenhouse effect. Theoretically, one hectare of jute plants can consume about 15 tons of CO_2 from atmosphere and release about 11 tons of oxygen in the 100 days of the jute-growing season. CO_2 absorption rate of jute is higher than trees.

The fast-growing seasonal crop jute reaches a height of 1.5 to 4.5 meters in a period of 4 to 5 months. The average dry stem production of jute ranges from 20-40 ton per hectare, annually. This difference with the production of the fastest growing wood plant which needs at least 10 to 14 years from plantation to harvest, and produces only 8 to 12 ton per hectare annually. As the biological efficiency of jute is much higher than that of wood plants, the use of jute instead of wood to make paper pulp will lower substantially the cost of production of pulp and paper and save forest resources.

There is a fertilizer value of removed jute leaves and it enhances the soil nutrients. Jute leaves are used as vegetables having nutritional and medicinal values. Jute sticks are used as a fuel and shelter in jute growing rural areas which contribute decreasing the use of wood in these applications.

The production flow of Jute associated with sowing, weeding/thinning, harvesting, defoliation and fermentation, fiber extraction, washing and drying. But only a small percentage of the farmers use seed treatment, fertilizers and herbicides/pesticides, which makes the processes before harvesting environmentally sound.

The manufacturing process of jute products involve several phases such as batching, softening with batching oil, carding, drawing, spinning, weaving and finishing. The use of mineral batching oils is being replaced with for specific use like packaging of Coffee.

Jute plant flourishes best in moist soil in a hot, humid climate. Seeds are hand-sewn, and plants mature in three months, often averaging a height of 10 to 12 feet (3 to 3.6 meters). Their light green leaves are arrow-shaped, and small yellow flowers bloom separately or in bunch. Jute is classified scientifically in the genus Cor chorus.

When the blossoms first start to shed, the plants are harvested. The cut stalks are sorted according to length and gathered into bundles. They are then placed in shallow pools of still water where they are permitted to ferment. When they become soft enough, the fibers are separated from the stalks and then hung on lines to dry. Fibers are sorted, graded, and baled for export after drying.

Burlap, low-grade twine, and many other products are made from jute. Its price is low and flexible. Jute is the second largest consumption of natural fibers to cotton in the world. India, Bangladesh and China are the leading producers.

The Golden Fiber jute has proved the huge popularity in the context of environmental issues. It is biodegradable and environment friendly. So, products combine with the soil after sustained use. In turn, it enriches the soil with organic substance and helps to grow better crops. It emits are non- toxic smoke during combustion process. No additional component or material remains after combustion. JRP (Jute Reinforced Plastic) is widely used to pack tea and fruits especially for its excellent 'breathing qualities'. It successfully packs garments, cement, fertilizers and other products as well. Geo-jute has been developed to control destruction on mountain slopes, canal banks and railway sidings. It also helps vegetation to grow - naturally. The perfect replacement of wood is jute.

Jute contributes manufacturing different type of products like fine silk, finished fabrics, useful furnishings to complicatedly designed oriental carpets, molded furniture, Wall Hangings, Swing Chairs, Flower Pot Holders, Tea Coasters, Mats, Blankets, Slippers, Shopping Bags, Bead Curtains, Dolls, Soft Luggage, Briefcases, Skirts, Jackets, Lamp Shades, Floor Runners, Panels, Boards and a whole lot more. They are inexpensive and aesthetic. Home and offices are perfect places of these products.

Green Products

Green products are the biodegradable environmentally friendly products. During using these products, there is no hazard in the earth and environment when it is released to the air, water or earth. These types of products decompose in a landfill quickly than similar non-biodegradable products. Biodegradable household cleaners, soaps, dishwasher detergents and laundry soaps are examples of this type of green product.

Types of Green Products

Product containing recycled goods in its construction, also considered green. Because, recycling reuses a material keeping it out of the landfill and saves on the environment when alternative materials are not manufactured and used for that component. If a product having reduced packaging in comparison to other similar products requiring less shipping room and can reduce carbon emissions during transport to market is also considered green.

How we will know that a product is 'Green'?

Green products should have labels showing that they are either biodegradable, have been made from recycled goods or were produced from an earth-friendly manufacturing process. So, green Products should have following health and/or environmental attributes:

- Promote good indoor air quality through reduced emissions of volatile organic compounds.
- Durable with little maintenance requirements.
- Incorporate recycled content (post-consumer and/or post-industrial).
- Recovered from existing or smashed buildings for reuse.
- Manufactured from natural and/or renewable resources.
- Require little energy to produce and transport materials.
- No CFCs, HCFCs or other ozone depleting substances in the product.
- No highly toxic compounds in the product as well as in the production process.
- Gained from local resources and manufacturers.
- "Sustainable Harvesting" practice exists in wood or bio-based products.
- Readily recyclable in a closed-loop recycling system.

Healthy Eating

In order to maintain good health, we should eat a healthy, balanced diet which can help us feel our best. It is not very difficult to maintain this food habit. For this, we should follow some food habits.

We should eat the right number of calories in order to remain active. Eating or drinking too much will put on weight. On the other hand, eating and drinking too little, will be the cause of weight loss. The average man needs around 2,500 calories a day (10,500 kilo joules). The average woman needs 2,000 calories (8,400 kilo joules). We should eat as per our required calories. Eating wide range of foods can ensure getting a balanced diet which is very essential for the nutrition of the body.

Starchy foods should make up around one third of the foods we eat. Starchy foods include potatoes, cereals, pasta, rice and bread. Choose wholegrain varieties or eat potatoes with their skins on. Because, they contain more fiber and can make us feel full for longer. We should include at least one starchy food with each main meal.

We should eat at least five portions of different types of fruit and vegetable per day. A glass of 100% sugar free fruit juice can count as one portion, and vegetables cooked into dishes also count. Eating banana in the breakfast is also good for health.

Fish is the source of protein and contains many vitamins and minerals. We should eat at least two portions a week, including at least one portion of oily fish. Oily fish is high in omega-3 fats, which help preventing heart disease. Examples of some oily fishes are salmon, mackerel, trout, herring, fresh tuna, sardines and pilchards. On the other hand there are some non-oily fishes like haddock, plaice, coley, cod, tinned tuna, skate and hake. Anyone who regularly eats a lot of fish should try to choose as wide a variety as possible.

Fat is one of the important factors in our diet. But, we should pay attention to the amount and type of fat which we are eating. There are two main types of fat: saturated and unsaturated. Eating too much saturated fat can increase the amount of cholesterol in the blood, which can increase the risk of developing heart disease. Saturated fat is found in many foods, such as hard cheese, cakes, biscuits, sausages, cream, butter, lard and pies. Try to cut down, and choose foods that contain unsaturated rather than saturated fats, such as vegetable oils, oily fish and avocados.

Instead of butter, lard or ghee we can use a just a small amount of vegetable oil or reduced-fat spread. During taking meat, choose lean cuts and cut off any visible fat. Always try to eat less saturated fat.

Foods and drinks having sugar are often high in energy (measured in kilojoules or calories), and could contribute to increase weight. They are also the reason of tooth decay, especially if eaten between meals.

We should try to avoid sugary fizzy drinks, cakes, biscuits and pastries, which contain added sugars. We can avoid this kind of sugar rather than sugars that are found naturally in foods such as fruit and milk.

Every day, we eat almost three-quarters of the salt in the food we buy, such as breakfast cereals, soups, breads and sauces. Eating too much salt can raise our blood pressure. People with high blood pressure are more likely to develop heart disease or have a stroke. More than 1.5g of salt per 100g means the food is high in salt. Adults and children over 11 should eat no more than 6g of salt per day. Younger children should have even less.

In order to maintain healthy weight, eating a healthy and balanced diet plays a vital role which is an important part of overall good health. Being overweight can lead to health conditions such as type-2 diabetes, certain cancers, heart disease and stroke. Being underweight could also affect our health. Adults need to lose weight, and need to eat fewer calories in order to control overweight. If you're trying to lose weight, aim to eat less and be more active. Eating a healthy, balanced diet is very helpful. Try to avoid foods contain high fat and sugar. We should practice eating plenty of fruit and vegetables.

In order to maintain weight loss or be a healthy weight, physical activity is very essential for us. Being active doesn't have to mean hours at the gym, we can find ways to fit more activity into our daily life. For example, try getting off the bus one stop early on the way home from work, and walking. Being physically active may help reduce the risk of heart disease, stroke and type-2 diabetes.

We should remember not to reward our self after being active with a treat that is high in energy. If we feel hungry after activity, we should choose foods or drinks containing lower calories but still filling.

Drinking every day about 1.2 liters of fluid can stop dehydration. This is in addition to the fluid we get from the food we eat. All non-alcoholic drinks count, but water, milk and fruit juices are the healthiest. We should avoid sugary soft and fizzy drinks that are high in added sugars and can be high in calories and bad for teeth.

Eating breakfast regularly can help people control their weight. A healthy breakfast is an important part of a balanced diet which provides some of the vitamins and minerals we need for good health. Whole meal cereal, with fruit sliced over the top is a tasty and nutritious breakfast.

Contribution of Green

What is Green?

Green is not only a color but also it has significant impact on earth and human being. When we think about green it reflects the beauty of this planet. Human being can make the nature of this planet greener which is very significant.

Why Green?

If we think profoundly, we can realize why green is necessary for our life as well as the planet. Basically, green is one of the wonderful creations of this planet. It keeps us alive. Animals like cow, horse etc. live on green grass. These animals are essential for our livelihood in terms of economic contribution.

Forests are the symbol of green and it is the factory of oxygen. When we leave carbon dioxide, trees take it and leave oxygen for us. So, we are benefited from green trees. Eco system is created by green nature. Green forests like mangrove can protect us from natural disastrous like cyclone, hurricane etc. Moreover, we enjoy the beauty of green when we share our time with green nature like forests and parks. Our mind becomes fresh and steady when we do meditation in a silent place like green forest or park. Also, green is the symbol of youth and enthusiasm. Sometimes, we use the word "Ever Green" in the sense of looking human being younger which is the inspiration for all.

Creation of green is totally under the control of human being in a sense that green can be made and destroyed by human being. We can make it and also destroy it. Initiating the tree plantation in front of open place of our house and office building can contribute to create a green and healthy environment. We can also create our building green by utilizing environmentally friendly technologies.